A MONOGRAPH ON

SHEATH BLIGHT DISEASE OF RICE

SHEATH BLIGHT DISEASE OF RICE

K. MANIBHUSHAN RAO
Centre for Advanced Studies in Botany
University of Madras, Guindy Campus
Madras 600 025, India

1995

DAYA PUBLISHING HOUSE
DELHI-110035

Front cover : Typical lesions on leaf sheaths of rice (L to R) : lesions on early stage, matured, coalesced and sclerotial initiation respectively.

Back cover : Sheath Blight (ShB) infected rice crop.
Note: whitish long irregular patches seen on the foliage.
(Courtesy: M/s Takeda Chemical Industries Ltd., Tokyo, Japan).

ISBN 81-7035-145-6

Published by : **Daya Publishing House**
1123/74, Deva Ram Park
Tri Nagar, Delhi - 110035.

Laser Typeset by : ShiKi Associates, New-Delhi - 110058

Printed at : **RAJ OFFSET PRINTERS**

Bound at : Book Wrap, Delhi - 110035

PRINTED IN INDIA

. . . . . .to my teacher
Professor Dr. T.S. Sadasivan

Acknowledgements

My sincere appreciation to the following authors, publishers and institutions who gave permission to reproduce plates, figures, tables and copyright © material used in this book. I thank CAB International, UK and the author Dr. J.E.M. Mordue for Fig. 1abcde; Dr. C.P. Dorairajan for Fig. 10ab; Dr. H. Hori , M/S Takeda Chemical Industries Ltd., Tokyo, Japan for Plate 1; M/S Nihon Nyhyaku Co. Ltd., Tokyo, Japan for Plate 2AB; Dr. K. Matsuura, Dr. T. Hirooka et al. and Dr. K.B. Johnson and copyright © 1986, 1989, 1994, The American Phytopathological Society, USA for Plates 9ABCDE, 10AB and 12ABCDEFG and Fig. 17 respectively; Prof. T. Hashiba for Table 1; Dr. Pat Crill for Tables 13 and 14; Dr. N. Sethunathan for Table 15; Mrs. T. Vasanthadevi et al., for Tables 16 and 17 (copyright © 1989) and Dr. S. Sreenivasaprasad for Figs. 11abc, 12 and Tables 19 and 21 for (copyright © 1990) reprinted by permission of Kluwer Academic Publishers, The Netherlands; copyright © 1989, 1993 by Eugen ulmer GmbH & Co., Stuttgatt for Plates 6, 7, 8; Figs. 8ab and Table 6; Prof. P. Vidyasekaran for Table 18; Prof. K.M. Rajan for Tables 23 and 24; copyright © 1982 by National Research Council of Canada for Figs. 4ab and Table 6; copyright © 1990, 1993 by Tropical Agriculture, Trinidad for Figs. 15 and 16 and Tables 25 and 26; Dr. S. Venkataraman for Table 30; Bibliographer, International Rice Research Institute, Manila, Philippines for Fig. 18. Finally, my erstwhile colleagues and a horde of research students for reproducing from several of the joint publications which are listed appropriately in the text.

Plate, figure and table numbers referred are from this book.

Detailed citations are available by consulting the bibliography.

Foreword

Sheath Blight is now one of the most serious diseases of rice. Control of sheath blight is essential for ensuring the realisation of the full potential of many high yielding rice varieties. It is in this context that the present monograph summarising the work done by Dr. K. Manibhushan Rao over a period of 30 years on this disease is a timely contribution. Dr. Manibhushan Rao has dealt with all aspects of the disease including varietal resistance and the genetic basis of resistance. The chapter on 'Challenges for the Future' would be found very useful by young scholars who are looking for problems of importance in rice production. I therefore hope the book will be read widely.

Madras
December 9, 1994

M.S. Swaminathan
M.S. Swaminathan Research Foundation
Third Cross Street, Institutional Area
Tharamani, Madras-600 113.

Preface

Rice in Asian countries is treated with reverence and its culture developed into an elaborate ritual. Even today, in South India especially in Tamil Nadu it is a practice in the farmers community to offer rice cooked in a new earthen pot to the Sun God on the 'Pongal' festival day. Though many of the beliefs and practices vanished, rice truly remains 'life itself' to the people of the world's densely populated regions especially in Asia. Incidentally more than 90 per cent of the world's rice production is harvested and consumed in Asia — a home to 59 per cent of the world's population. To meet the projected rice in demand, we need about 758 million tonnes(mt) by 2025—i.e., 70 per cent more rice than at the present consumption - just to keep the population growth.

With the ever increasing population and the escalating demand for food, scientists are toiling hard to enhance the production level. Before the release of the high yielding IR 8, the 'miracle rice' in the early 1960s from International Rice Research Institute (IRRI), Philippines and subsequent introduction of IR-type semi-dwarfs, sheath blight (ShB) was not a problem in rice production in Asia, especially South East Asia and was considered as a mere scientific curiosity, but, in recent years, it has become a potential threat to rice cultivation in India. It is now considered as one of the principal diseases as it causes more losses than any other fungal diseases of rice.

The recommendations of the 'green revolution' namely, intensive cultural practices especially close planting and heavier application of nitrogenous fertilizers on improved varieties, leading to heavy tillering and good growth, unfortunately favour the disease incidence. Further, the progressive and wealthy farmers also create an ideal situation for the spread of *Rhizoctonia solani* (syn. *Corticium sasakii*) by using a single glamorous variety and also by applying high doses of N fertilizer (about 200 kg N/ha and above). This would be an excellent precondition for the spread of this disease like a wild-fire. Thus, it is appropriately called the 'rich man's disease'.

Against this background I have initiated studies on ShB in 1976 after spending a sabbatical year at IRRI, Philippines. The significant developments in 1980s make a complete review almost impossible. However, this monographic study was aimed to update advances of the last two decades and the current perspectives and challenges as well as specific problems that have surfaced.

I have studied the ShB of rice from different angles with the cooperation of my research students, with an emphasis on biocontrol measures. The peculiar versatility of the pathogen tended to defeat attempts to breed for resistance and practically no highly resistant or immune source was identified. Further, the subterranean nature of the pathogen again renders the chemical control unsuccessful, and use of induced alternatives is envisaged. I have also included pertinent available information and material from other rice workers for which the individual acknowledgements were made separately. Particular attention is given to inclusion of the latest available information, a factor, I deem important in view of the tremendous work on ShB is initiated on biocontrol in recent years. I intend this monograph for world wide usage by students, researchers and extension workers in rice pathology.

I thank Dr. Roy Johnson, Plant Pathologist, AFRC Institute of Plant Science Research, Cambridge Laboratory, Norich, U.K. and Prof. A.S. Rao, Professor Emeritus, Nagarjuna University, Nagarjunanagar, India for critically going through the draft of the MS and for their valuable comments and suggestions which proved extensively valuable to me while writing this book. I own complete responsibility for any errors, misinterpretations and deficiencies that may remain. I shall always welcome any further comments and suggestions from other rice researchers.

I have great pleasure in thanking Professor M.S. Swaminathan, Chairman, M.S. Swaminathan Research Foundation, Institutional Area, Tharamani, Madras and former Director-General, International Rice Research Institute, Philippines who has been the main brain behind "green revolution" in this country and also writing the foreword for this monograph.

I wish to record my grateful appreciation to the University Grants Commission (UGC) and Indian Council of Agricultural Research (ICAR), New Delhi, India for extending continued financial support to my major research projects on rice diseases over these years. I appreciate the cooperation of M/S Daya Publishing House, Delhi for bringing out this monograph at short notice. I thank the University of Madras for its support and to Prof. A. Mahadevan, Director, CAS in Botany for facilities. Lastly, the cooperation and help from my wife Mrs. Annapurna in more ways than one is gratefully acknowledged.

Madras **K. Manibhushan Rao**

January 1, 1995

Contents

Standard Abbreviations

are(s)	:	100 sq. metres
°C	:	degree Celsius
cm	:	centimetre(s)
cv(s)	:	cultivar(s)
CFU	:	colony forming units
DAS	:	days after sowing
DAT	:	days after transplanting
dia	:	diameter
g	:	gram(s)
g^{-1}	:	per gram
h	:	hour(s)
ha	:	hectare (s)
ha^{-1}	:	per hectare
kg	:	kilogram(s)
HRLH	:	highest relative lesion height
l	:	litre(s)
m	:	metre(s)
ml^{-1}	:	per millilitre
µ	:	micron($x10^{-6}$)
m^2	:	square metre(s)
mg	:	milligram(s)
min	:	minute(s)
ml	:	millilitre(s)
mw	:	molecular weight
N	:	nitrogen
p.(pp.)	:	page(s)
pH	:	hydrogen ion concentration
ppm	:	parts per million
RH	:	relative humidity
sec.	:	second(s)

contd. . . .

sp.; spp.	:	species
t	:	metric ton(s)
wk	:	week
±	:	with or without
%	:	percent(age)
/	:	per
σ	:	sigma
α	:	alpha
ß	:	beta
γ	:	gamma
p	:	para
o	:	ortho
m	:	meta
r	:	correlation
y	:	observation

History and Distribution

The sheath blight (ShB) or the banded blight/sclerotial disease was first reported in Japan by Miyake (1910) who named the incitant as *Sclerotium irregulare*. Later, Sawada (1912) found that the pathogen was identical with *Hypochnus sasakii*. Subsequently, Reinking (1918) and Palo (1926) reported similar disease in the Philippines and considered the fungus to belong to the *Rhizoctonia solani* group. Further, Park and Bertus (1932) reported the disease in Sri Lanka and identified the pathogen as *Rhizoctonia solani* Kühn,* while Wei (1934) noted the disease in China and thereafter it has been found in many Asian countries. Ryker and Gooch (1938) from the US described sheath spot disease, with symptoms which are very similar to ShB. In the southern USA it is recognized as one of the economically important rice diseases (Templeton and Johnston, 1969; Rush, 1971; Atkins, 1974; Boyettee and Lee, 1979). Reports of ShB occurrence from Nigeria (Awoderu, 1974; Imolehin, 1989) are available.

In India, Paracer and Chahal (1963) were the first to record the rice ShB fungus as *R. solani* from Gurudaspur, Punjab and later from Uttar Pradesh (Kohli, 1966). Further, Saksena and Chaubey (1972) described it as *Thanatephorus cucumeris* causing banded blight disease. Subsequently, the occurrence of ShB on rice was reported from several states in India viz. Andhra Pradesh, Kerala, Tamil Nadu, Uttar Pradesh, Punjab, Orissa, etc. where rice is predominantly grown.

Initially, rice ShB was reported as a minor disease, a mere scientific curiosity but now distributed widely in rice growing countries from the tropics to the temperate regions and is now positioned next to blast in economic importance (Rush, 1971; Yamada, 1986). The importance of this disease is still tending to increase, particularly in East Asian countries (Manibhushanrao et al., 1979), often as a result of the introduction of semi-dwarf modern cvs. and increased crop intensity (Mew, 1988).

In India, during the early 1970's the cultural practices of 'green revolution' especially close planting and heavier application of N-fertilizers on semi-dwarf, high yielding, improved cvs. with compact growth and large number of tillers, resulted in an increase in humidity of the plant layer thereby favouring the spread of the disease. In the recent past, in a seed farm at Adoor (Kerala state), India the ShB epidemic was so severe that the fields were abandoned as even hay could not be harvested. This has caused great concern to farmers in south India as the disease outbreak may occur at any time, place and magnitude under favourable conditions.

Further, extensive studies of ShB have been made in Japan, Korea, Philippines and India with an emphasis on varietal resistance and chemical and biological control.

---

* Most voluminous literature is carried under this name and hence it is felt appropriate to retain *R. solani* although a few publication of the author as well as other rice workers appeared with *Corticium sasakii* and *T. cucumeris*.

Comprehensive reviews (Manibhushanrao et al., 1979; Gangopadhyay and Chakrabarti, 1982; Lee and Rush, 1983; Manibhushanrao, 1989) highlighted the importance of ShB as a major disease of rice. Subsequently, an extensive coverage on rice ShB was made (Gangopadhyay, 1983; Ou, 1985; Dath, 1990; Manibhushanrao, 1990, Dasgupta, 1992).

The incitant *R. solani* causes many types of diseases on an array of economically important crops (Baker, 1970) and a detailed review on biology and pathology of *R. solani* is available (Parmeter, 1970). The fungus is well known for its versatility as plant pathogen, a parasite, a symbiont or a facultative saprophyte and has a wide host range from either a semi-arid or even an aquatic plant. Neither the former nor the latter plant escapes when challenged with an appropriate strain. It causes several diseases of crops of economic importance and, perhaps, damping-off of seedlings of most crops is the principal disease of *R. solani*. Here, the discussions are mostly confined to rice ShB only. However, a few other aspects relevant to *R. solani* are discussed.

Plate 1 ABCDEFG : Showing typical sheath blight lesions on leaf sheaths. (A) lesions on leaf sheaths in early stages, (B) at later stages, (C) formation of hymenia with basidiospores on leaf sheaths, appear as whitish powder, which cover leaf blades and leaf sheaths, (D) formation of whitish sclerotia in early stage, (E) mature sclerotia are brown in colour, (F) lesions on leaf blades and (G) lesions spread rapidly to the upper plant parts (Hori, 1971).

Plate 2 AB : (A) Germinated sclerotia attached to the sheath (enlarged circles) with whitish mycelia (black background). (B) well developed lesions and sclerotium (enlarged circle) detached from the sheath. (Courtesy : Nihon Nyhyaku Co., Ltd. Tokyo, Japan).

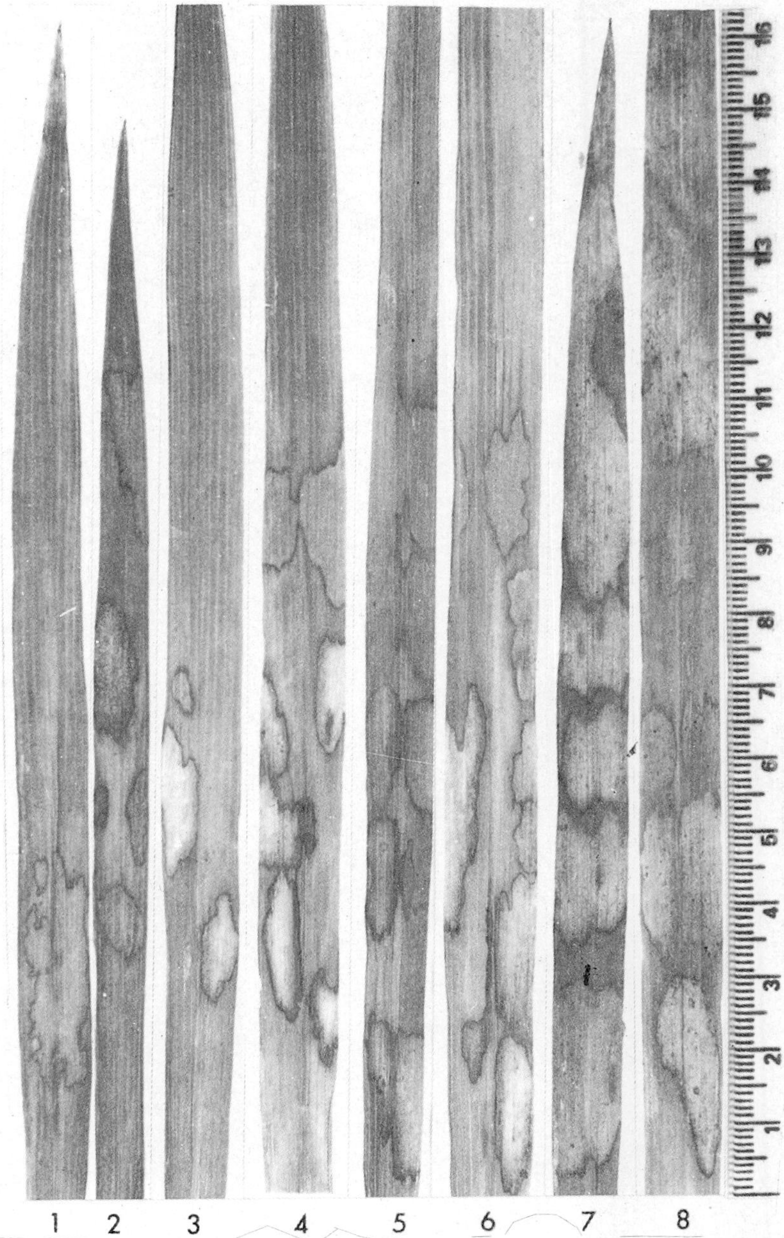

Plate 3 : Showing typical sheath blight symptoms on leaves of field grown susceptible cv. BPT 1235. 1 and 2 show lesions on initial stages of infection; 3-6 show enlarged, coalesced lesions irregular dark brown with greyish white centres and 7 and 8 show advanced stages of infection where leaves are completely discoloured and withered by the coalescence of lesions (Manian, 1982).

Symptomatology

This disease causes lesions/spots on seedlings, leaf sheaths and on leaves. Lesions initially appear as circular, oblong or ellipsoidal, greenish-gray, water-soaked spots, about 1 cm in length. The spots soon enlarge to 2-3 cm length, with irregular dark brown margins, while the centre is bleached to greyish white. Under field conditions, the spots are usually seen on the leaf sheaths (Plate 1 A,B). Basidiospores appear as whitish powder (C) on leaf blades and leaf sheaths. The presence of several large lesions on the leaf sheaths and their coalesence cause death to the entire leaf and in severe cases all the leaves of a plant become blighted and withered. Small whitish round sclerotia appear in early stage (D) while the mature ones turn brownish in colour (E). Sometimes the lesions on leaf blades are mistaken for leaf blast (F). The lesions spread rapidly upwards under high humidity and temperature conditions (G).

The sclerotia attached to rice plant start germinating on the sheath (Plate 2A, enlarged circles). Mycelia developed from sclerotia penetrate into sheath cells. Mature sclerotium formed on well developed lesions, fall down (Plate 2B, enlarged circle) to the ground and function as inoculum source in the next season. Typical ShB lesions and symptoms are shown on leaf blades (susceptible cv. BPT 1235) under field conditions (Plate 3). The fungus is over grown on the leaf tips when the rice seedlings are inoculated artificially and in advanced stages become black and withered.

The size, colour of the lesions on the leaf sheaths will depend on the susceptibility of the host as well as environmental conditions. Under favourable humid conditions the mycelium of the fungus from the enlarged lesions can spread to a considerable distance up to leaf tips. Sclerotia are seen prominently near the mature lesions and fall on the water surface and incite infection on the leaf sheaths near the water line.

Disease Development

Though ShB is primarily soil borne, it is seed borne when infection reaches the panicles from the bottom sheaths. The fungus can live even for 16 and 21 months under dry and moist conditions respectively in soil and on debris remaining in the field. During the puddling and transplanting, the sclerotia that float on water may be carried away to other neighbouring fields and infect the rice plants on contact at water level. After harvest of the crop, white crusts of fungal growth appear on the surface soil. It is within them that millions of basidiospores are produced. These are carried by wind, causing aerial blight on host plants. Hashiba et al. (1983) studied the mode of upward or vertical development of ShB.

Natural infection usually occurs at the seedling, tillering and boot-leaf stages of the rice crop. The disease development is most rapid at heading stage (Yu et al., 1980; Roy, 1982) in irrigated and rainfed ecosystem (Greenland and Dutta, Mew et al. 1986). However, the disease is more prevalent in dapong nurseries, which are regularly adopted by farmers to cut down the area and cost of raising plants. It is also most prominent where the crop is grown under intensive production system.

Early maturing cvs. are more susceptible than the late maturing ones (Ou et al., 1973). The mycelial development of the fungus is closely related to the growth stage of the rice plant and to climatic factors. Maximum mycelial development occurred 10-d before and after heading (Peng et al., 1965). Kozaka (1961) described two phases in the disease development-upward, occurring only after the heading stage and the more rapid horizontal one developed in the maximum tillering stage. Despite the favourable environmental conditions, the infection process is primarily dependent on the number of sclerotia in the immediate vicinity of the plant and its susceptibility. Older plants are more susceptible and prone to the disease than the seedlings (Kozaka, 1961). Further, Kim et al., (1985) studied the relationship between the inoculation time of sclerotium and disease development of ShB.

Kannaiyan and Prasad (1981a) reported that rice chaff reduced seedling infection by ShB fungus. However, Dath (1985a) claimed that infected chaff and rachilla are the potential sources of infection to rice plants with ShB infection in the double cropped areas. To reduce the inoculum potential for ShB, the infected chaff and rachilla should be destroyed from the vicinity of the rice crop. Destruction of weed grasses, stubbles (Valdez, 1955) infected rice straw (Lee and Cournteny, 1982) were suggested earlier to limit the disease spread. Sharma and Teng (1990a) studied ShB development and yield loss on four different growth stages (tillering, panicle initiation, booting and flowering) of the rice cvs. (IR 42 and IR 72) and found that rice is more susceptible at booting and flowering than at tillering and panicle initital stages. Yield ranged almost same for both cvs. from 3.7 - 5.6 t/ha while yield loss due to ShB may occur at any stage but is higher when infection occurs at booting or flowering.

Yield Losses

Both in tropic as well as under temperate conditions the yield losses due to ShB ranges from 20 to 50% have been reported (Miruta, 1956; Hori, 1969; Wu, 1971; Kozaka, 1975; Tasi, 1975; Boyatte and Lee, 1979; Arunyanart et al., 1984; Hashiba, 1984; Ahn and Mew, 1986; Rajan, 1987). Under severe conditions more than 70% loss in yield was estimated (Baby, 1992). Earlier, Ono (1953) correlated the relationship between the plant parts infected by ShB and crop loss as 15% between 3rd and 4th leaf, 20% between 2nd and 3rd leaf and 40% between flag leaf and grain filling stage.

A close correlation could be obtained between the development of ShB and yield (Hori, 1969; Yu, 1977; Kannaiyan and Prasad, 1978a; Hashiba, 1984a,b). Timely application of fungicides on infected plants of 5, 50 and 100% of the hills increased the yield by 1.6, 7.1 and 10.1% respectively. Further, Ou and Bandong (1976) reported 7.5-22.7% yield loss in the low N plots and 8.6-23.7% in the high N plots for a susceptible cv. while for a moderately resistant cv. a loss 0.4-8.8% and 2.5-13.2% was observed. In China, despite all efforts of the management the crop loss due to ShB is about 10% but in Korea the losses were very heavy (Crill et al., 1981). In Vietnam, ShB has caused total crop failure in recent years while in other countries as high as 50% crop reduction was reported (Ou, 1985). Generally, in the tropics the highest crop losses occur. Tsai (1976) measured yield losses by inoculating plants at different stages of their development while Hashioka et al., (1983) estimated based on height of lesions and per cent affected hills. Although ShB incidence markedly increased in Sri Lanka, the yield losses were not substantial (Wickremasingha and Mithrasena, 1988).

Hashiba (1985b) reported the yield losses in ShB through a forecasting model while Hashiba and Ijiri (1989) estimated the actual of yield loss in the rice field of ShB incidence by the forecasting model BLIGHTAS. This was based on the ratio of the height of lesions to the plant height (i.e. vertical development of the disease and percentage of the number of diseased hills i.e. horizontal development of the disease). In this system, the yield loss of fully ripened kernels in kg/10 ares (L) is shown as follows: $L = (41.31X - 826.2)\ A/1000$; here, the yield loss actually observed in paddy fields from 1976-1980 was agreed well with the estimated one (Table 1). This seems to be the first computerized forecasting system for the rice ShB and the authors claim that this system can be applied for the development of new strategies to control ShB. Further, BLIGHTAS has the advantage of predicting the disease incidence and subsequent yield loss within 15-20 min using a compatible microcomputer so that protective measures can be taken in much advance of disease epidemic progress.

Table 1: Comparison between the actual yield loss observed in fields from 1976 to 1980 and the yield loss estimated by the forecasting model (Hashiba and Ijiri, 1989).

Year		X	A	D	L
1976	1	45.0%	17.0%	6.9%	13.3%
	2	32.3	22.6	8.4	21.5
1977	1	48.2	23.0	10.5	26.8
	2	51.6	31.7	16.7	42.6
1978	1	37.1	14.0	3.9	9.9
	2	34.0	9.3	2.2	5.5
1979	1	46.0	16.0	6.7	17.2
	2	43.7	19.3	7.6	19.4
1980	1	47.0	25.6	11.2	26.2
	2	51.0	26.4	13.7	34.8

1 : Observed value
2 : Estimated value
X : Relative height of the uppermost lesions to the plant height (%)
A : Percentage of diseased hills (%)
D : Disease incidence (%)
L : Yield loss (kg/10 ares)

Extensive studies on yield losses due to ShB by evaluating disease development in the field have been carried (Basu et al., 1992; Pillai, 1993; Yokoyama and Okuhara, 1992) from time to time. However, the method of estimation of yield loss is viewed and the subsequent developed computerized forecasting systems (Ohata and Shimada, 1990, 1992; Ozawa and Makano, 1990) have to be fully established.

Of late, ShB is considered internationally important disease as it is a foremost rice disease widely distributed in irrigated and rainfed lowland ecosystem especially in Asia. An efficient disease control strategy is required to minimize the yield losses.

Causal Organism

Rhizoctonia solani Kühn : (Class : Deuteromyceteae; Order : Mycelia Sterilia).

Rhizoctonia solani Kühn, in "Die Krankheiten der kulturgewachsen, ihre ursachen und ihre Verhiitung Gustav". Bosselmann, Berlin, 1858.

Rhizoctonia oryzae Ryker and Gooch, Phytopathology **38** : 238, 1938.

Thanatephorus cucumeris (Frank) Donk, Reinwardtia **3** : 376, 1956.

Waitea circinata Warcup and Talbot, Trans. Br. mycol. Soc. **45** : 503, 1962.

Rhizoctonia solani Kühn f. sp. *sasakii* (Exner) Ahuja and Payak, Curr. Sci. **32** : 328-329, 1963.

Thanatephorus sasakii (Shirai) Tu and Kimbrough. Bot. Gaz. **139** : 147, 1978.

Nomenclature

The nomenclature of the rice ShB fungus is under dispute and it has been described in several synonyms such as *Corticium sasakii* (Shirai) Matsumoto (Matsumoto, 1934; Ou, 1985); *Sclerotium irregulare* (Miyake, 1910); *Hypochnus sasakii* Shirai (Shirai, 1906 and Sawada, 1912); *Rhizoctonia solani* Kühn (Reinking, 1928; Palo, 1926; Park and Bertus, 1932); *Corticium solani* (Prill and Delacr) Bourd. and Gtalz.; *C. vagum* Berk and Curt. (Gadd and Bertus, 1928); *Pellicularia filamentosa* (Pat.) Rogers f. *sasakii* and *P. sasakii* (Shirai) S. Ito (Rogers, 1943). However, Rogers' nomenclature was rejected by Venkatnarayan (1949) and Talbot (1965) who considered the genus *Pellicularia* Cooke as a *nomen confusum*.

The perfect state (basidial state) of the fungus was reported in India by Singh and Pavgi (1969), subsequently, Saksena and Chaubey (1972) and in Philippines by Ou et al. (1973). The basidial state appears as a white powdery or frosty layer on healthy leaf sheaths and occasionally on leaves near the infected tissues under extremely moist conditions. Basidia and basidiospores measured those resembled that of *Thanatephorus cucumeris* (Frank) Donk. Further, Talbot (1970) considered *T. cucumeris* to be the correct name and treated it as an aggregate of species including the ShB fungus. According to him both *Corticium* and *Hypochnus* are unacceptable for both taxonomic as well as nomenclatural view points. However, Tu and Kimbrough (1978) have treated the rice fungus as a species separate from *T. cucumeris* and recommended the name as *T. sasakii* (Shirai) Tu and Kimbrough.

Taxonomy

Earlier Endo (1927) examined the type specimen of the fungus from the Philippines (Palo, 1926) and found it closer to *Hypochnus sasakii* than *R. solani*, while Matsumoto et al. (1932; 1933) and Matsumoto (1934) made some observations on the taxonomy of the fungus *H. sasakii* obtained from various hosts and concluded that the morphological differences in the teleomorph alone are not sufficient to classify *H. sasakii* as a distinct species, and for the vegetative characters it could not be considered as *C. vagum*. However, they considered *C. sasakii* as the most acceptable. In this context, the causal organism, the incitant of ShB disease, described as *R. solani* in most of the voluminous literature that has appeared, although a number of publications of the author as well as others carried with *C. sasakii* and *T. cucumeris*. However, no evidence is lacking for the adoption of *R. solani*, that remains controversial. It is pertinent to consider the name referred to the organism in this text as the 'sheath blight fungus' (ShB fungus) as suggested by Ou (1985).

Morphology

The fungus produces two main different types of mycelia on the host: The straight, non-infectious runner hyphae (4.2-10.5 (7.0) µm) and the short, swollen, multibranched, infectious lobate hyphae (6.7-10.0 (10) µm) produced from the former and give rise to the penetration pegs. The mycelium is hyaline when young, turn yellowish-brown when old. The lobate mycelium penetrates the host tissue through infection cushion and produces lesions. Generally, the runner hyphae covers most part of the infected plant (stem and leaf blades) but the lobate ones confine to the lesions only. A third group of mycelium consists of monilioid cells involved in the formation of sclerotia are formed in *in vitro* culture of the fungus especially on the walls of petridishes or test tubes.

The summary of teleomorph features described by various workers for *Corticium vagum* (Burt, 1926), *Hypochnus sasakii* (Sawada, 1912); Ikata and Hitomi, 1930; Matsumoto et al., 1932) and for *Thanatephorus cucumeris* (Mordue, 1974) is given in Table 2 (Manibhushanrao, 1990). Detailed morphology of vertical section of sclerotium, sclerotial cells, hyphae from aerial mycelium, monilioid cells from PDA culture of *Solanum tuberosum* and basidia and basidiospores (Fig. 1) on *Newbouldia* leaves is available (Mordue, 1974).

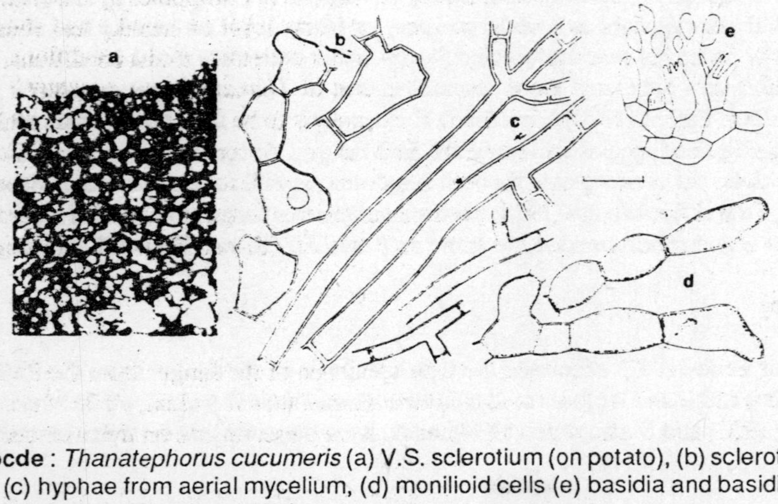

Fig. 1 abcde : *Thanatephorus cucumeris* (a) V.S. sclerotium (on potato), (b) sclerotial cells, (c) hyphae from aerial mycelium, (d) monilioid cells (e) basidia and basidiospores Bar : 20µm (Mordue, 1974).

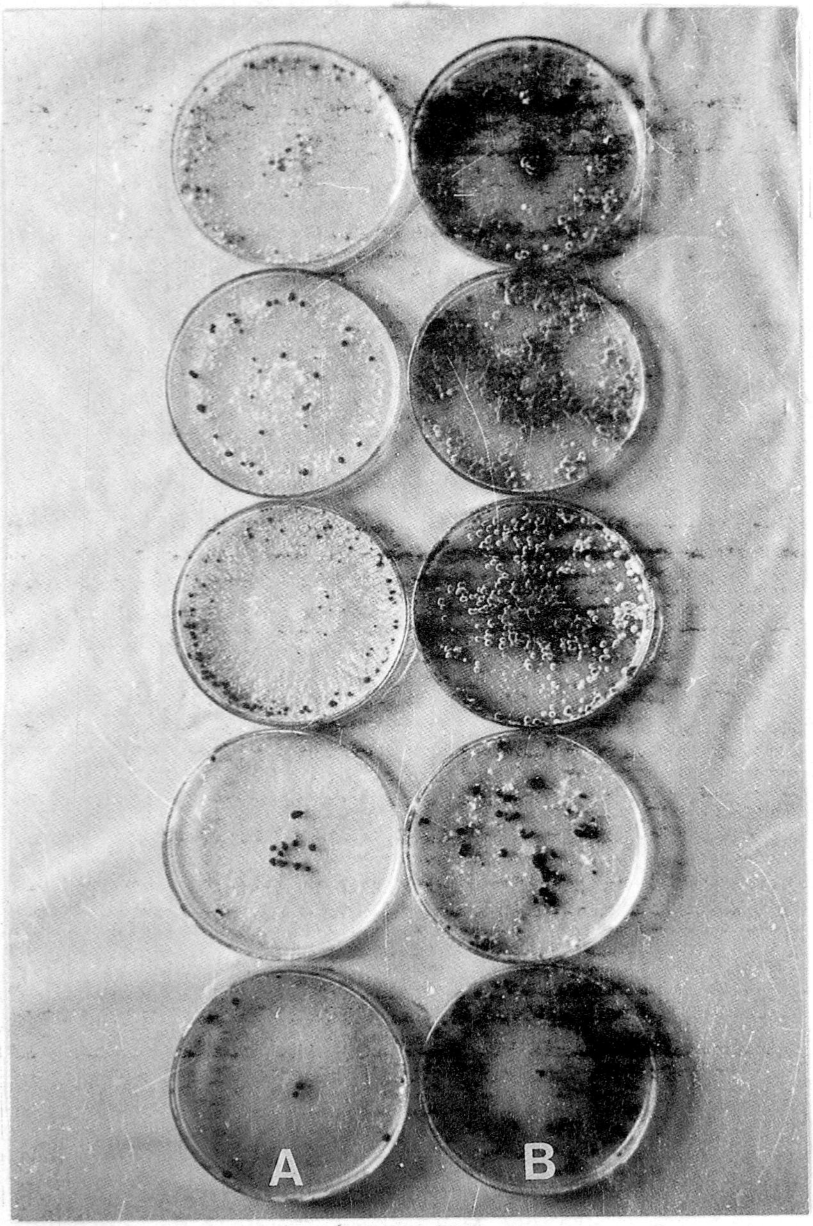

Plate 4 AB : *R. solani* isolates showing range of cultural characteristics, type and pattern of sclerotial formation on two different media. (A) top to bottom R_1 to R_5 isolates on PDA medium and (B) Czapek (Dox) agar medium (Manian, 1982).

Table 2: The summary of the teleomorph features (Manibhushanrao, 1990)

Fungus / Author	Range (µm) Basidia	Sterigmata	Basidiospore
Corticium vagum			
Burt (1926)	10-20 x 7.5-11	6-16	8-14 x 4-6
Hypochnus sasakii			
Sawada (1912)	10-15 x 7-9	4.5-7 x 2.3	8-11 x 5-6.5
Ikata and Hitomi (1930)	5.6-12 x 5-8	2.5-8.4 x 0.8 x 2.8	6.5-8.4 x 4.8-8.4
Matsumoto et al., (1932)	10-16 x 8-9	5.8 x 2.2-2.7	6-10 x 4-7
Thanatephorus cucumeris			
Mordue (1974)	10-25 x 6-19	5.5-36.5	6-14 x 4-8

The teleomorph has been observed in rice seedlings in an upland nursery when inoculated with a mycelial culture of ShB fungus (IRRI, 1973). No further reports are available on isolates from rice.

When the mycelium is 6 to 8-d old, it starts producing sclerotia. They are round in shape, flattened below, 1-5 mm dia., composed of compact masses of hyphal cells and brown to dark brown in colour. Average mature sclerotium in culture measures up to 5 mm but many units coalesce to form a large globose mass under favourable conditions.

The browning is completed after 40 h by which time the cells in the outer layers start to become empty. The sclerotia are initially dense and sink in water but after 15-ds of growth the outer layers diminish, become light and float in water. Hashiba and Staples (1976) reported the metabolic changes occurring in a developing sclerotia.

The morphology of the sclerotia and the mycelia varies greatly among ShB fungus isolates (Plate 4 AB). Three morphological groups were identified among the several isolates isolated on Richard's agar (Kohil, 1966). In Taiwan, 20 strains were detected among 310 isolates of which 50% fell in three groups (Tu, 1967). Further, Tu (1968) reported the physiological variation in seven strains of *C. sasakii*, although closely related and compatible, differ in their pathogenicity. From cross-inoculation studies, Saikia and Roy (1975b) concluded that the rice and turmeric isolates did not differ either in pathogenicity or morphological characteristics.

Great variation in colour and type of mycelium, size, number and type of sclerotia were observed when cultured on agar medium. Variation in cultural characteristics of the sclerotia was also observed in field isolates (Kernkamp, 1952; Sherwood, 1969; Parmeter and Whitney, 1970). Variations in the survival of sclerotia was attributed to soil microorganisms (Naicki and Ui, 1975; Mew and Rosalis, 1984).

Based on the number of nuclei in each hyphal cells Hashioka and Makino (1969) attempted to separate various species of the fungus. They found that *C. sasakii* has mostly 6-8 nuclei, *C. solani* 4-9, *C. gramineum* 1-3, *C. rolfsii* more than 10 and *R. oryzae* 3-5. Further, Tsai (1973) and Haque (1975) have studied variation in cultural characteristics, and pathogenicity of 40 single basidiospore isolates and 50 field isolates respectively from IRRI, Philippines.

Kamjaipai and Giatgong (1971) reported three isolates with average of 5 or 6 nuclei from Thailand and one isolate with 2 nuclei from Laos. Inagaki and Makino (1974) studied 12 isolates from rice from different localities and found 6-8 nuclei in most cells (average 7.3) with slightly more in isolates from the US and Hyogo (Japan) and fewer in isolates from Thailand and Wakayama (Japan). The range was from 1-31 nuclei per cell.

Endo et al. (1974a, b, c) made extensive comparative studies on the two isolates of *C. sasakii* from rice and crimson clover causing ShB and summer blight, respectively, and *R. solani* from crimson clover also causing summer blight. The isolates of *C. sasakii* from rice and crimson clover differed in their pathogenicity to some forage grasses, legumes and rice. They were, however, similar in mycelial growth, sclerotial formation, colour of mycelial mat, optimum temperature for growth on several media, growth in mixed cultures and shape of sclerotia. *R. solani* from crimson clover was remarkably different from both the *C. sasakii* isolates in sclerotial morphology and cultural characters. The former two isolates differed from *R. solani* in the reaction type produced on the sheaths of the rice plant. Although morphologically indistinguishable, they were considered differently as physiological strains of *C. sasakii* since they differed in their pathogenicity. This was further confirmed by the results obtained from the studies on hyphal anastomosis between the strains of *C. sasakii* (Endo et al., 1976). Thus it was also demonstrated from these experiments that *R. solani* and *C. sasakii* are not closely related.

Cultural types

Watanabe and Matsuda (1966) established six cultural types (IA, IB, II, IIIA, IIIB and IIIC) based on their variations in growth rate, production of sclerotia and general colony appearance. Nonaka (1977) isolated culture type IIIB of *R. solani* which caused brown ShB lesions in rice. Vijayan and Nair (1981) and Thakur et al. (1992) studied the anastomosis grouping of rice isolates while Das and Balakrishnan (1978) reported in cowpea isolates of *R. solani*. On the basis of anastomosing ability, most isolates of *R. solani* fall in five anastomosis groups (AGs) namely AG-1, AG-2 AG-3, AG-4 and AG-5 (Anderson, 1982). *R. solani*, the incitant of the rice ShB is referred to be the AG-1 (Parmeter et al., 1969; Sherwood, 1969; Ogoshi, 1972, 1975, 1976 and Nonaka et al., 1979). Yamaguchi et al. (1982) isolated anastomosis group AG-4 from rice plants in addition to AG-1 and AG-2 in Indonesia. Kim et al. (1988) studied a total of 520 isolates of *R. solani* from sclerotia overwintered in paddy fields and classified all into AG-1 group. Kim and Yoshino (1988) studied the relation to locality and cultural age of the AG1-1 isolates. Most workers referred AG-1 group for rice isolates besides AG-2 and AG-4.

It appears that *R. solani* is not a single species but composed of groups based on hyphal anastomosis patterns. Earlier, Talbot (1970) recognised four main groups and in one 'praticola group', 'H' factors were shown to control heterokaryosis while Naiki and Kanon (1978) classified into five anastomosis groups according to their virulence. Ogashi et al. (1979) reported group specificity among the isolate while Roy (1979) identified two isolates with 'R' and 'B' differing in their pathogenicity to rice cvs. More detailed information about the anastomosis groups and strain variation in *R. solani* is needed before differentials can be selected on priority basis.

Cell and molecular biology

For assessing genetic variation in fungi one of the direct approaches will be the study of molecular markers. The gene coding for ribosomal RNA (*r*-RNA) are well suited as markers as they evolve more slowly than many other parts of the genome, yet are large enough for evolutionary comparison. Recently, Vilgalys and Gonzalez (1990) surveyed intraspecific groups within the complex of 87 isolates of *R. solani* for restriction fragment length polymorphism (RFLP) within nuclear encoded *r*-DNA genes. The genemoic DNAs were studied for RFLPs by digestion with the restriction enzymes *Eco* R1 and *Bam* H1 and analyzed by southern blotting against *r*-DNA probe originating from anastomosis group 4. Each of the 15 intraspecific groups of *R. solani* are characterized by one or more RFLPs. Levels of intragroup RFLP variation were within the ranges of RFLP variation of different AGs and their subgroups. Isolates from AGs 3,4,7,8 and B1 all possessed a single invariant RFLP unique to each intraspecific group. A relatively low level (AG-I-1B, AG-I-1C, AG-2-2 and AG-9) and high level (AG-1A, AG-2-1, AG-5 and AG-6) diversity was observed, however, no obvious relationship was noticed between levels of RFLP within intraspecific groups and variation of other biological factors such as host range, morphological variation etc. Thus, the pattern of RFLP variation observed is most likely the result of genetic divergence within and among interspecific groups.

Earlier studies measuring heterologous DNA/DNA hybridization showed that extensive variation was present within many AGs and low sequence complementarity among most intraspecific groups (Vilgalys, 1988). Further, DNA hybridization experiments (Kuninaga and Yokosawa 1984a, b; Kuninaga et al. 1979) revealed the existence of intraspecific groups in AG-4 and AG-6. They introduced the term 'homogenous group' for differentiating isolates of these AGs based on their DNA similarity. However, this terminology has not resolved the problem of interspecific taxonomy in the complex of *R. solani*. The non-homogenicity of RFLPs among anastomosis groups of this fungus also supported the common belief that they represent independent evolutionary units (Anderson, 1982; Ogoshi, 1975). Here, the higher levels of variation among RFLPs within some interspecific groups might suggest the availability of additional divergent groups.

Physiology

Isolates of ShB fungus from rice appears to be adopted to the adverse conditions of the environment. Its diversity in form is reflected in the diversity in its function also. It can tolerate a wide range of pH (2.5-7.5) the optimum being 6.5. The temperature range is also wide (10-42°C) and it possesses the ability to utilize a wide spectrum of carbon and nitrogen sources. It is also capable of producing phytotoxic secondary metabolites. Philip and Joshi (1980) reported *in vitro* toxicity of boron to *Corticium sasakii*. Other workers studied nutritional physiology (Manian, 1982), electrophoretic patterns of proteins and enzymes (Zuber and Manibhushanrao, 1982) and toxin production (Ramalingam, 1983b).

Temperature

Hemmi and Yokogi (1927) reported that the optimum temperature for mycelial growth was 30°C with a maximum of 40-45°C and a minimum of 10°C, where little or no growth was observed. Subsequently, Matsumoto et al. (1933) reported the optimum range as 28-31°C, while Tu (1968) observed it as 28°C. Santos (1970) identified four temperature groups among the isolates. No correlation could be obtained between the optimum temperature and the rate of growth among 40 single basidiospore isolates (Tsai, 1973). Among several field isolates tested Haque (1975) reported that 2 out of 25 isolates were sensitive to high temperatures, while one was sensitive to low temperature. Hashiba et al. (1974) observed that isolates from high temperature regions grew well on PDA at 35°C and poorly at 12°C, and vice versa in the case of isolates from low temperature regions (IRRI, 1971). However, Kim et al. (1988) did not get any significant differences in their maximum and minimum temperatures from the mycelial growth of the isolates of *R. solani* from overwintered sclerotia in Korea. Perhaps the local regions of the isolates did not relate to specific thermal reactions for mycelial growth because the geographical location of Korea belongs to the same temperature zone.

Hashioka and Makino (1969) found that *R. oryzae* had an optimum temperature requirement of 25°-30°C and could grow well even at 32-35°C, while in the case of *C. sasakii* they were 25-30°C and 20-25°C, for *C. solani* 25-27°C and 27-30°C. Inoue and Uchino (1963) reported that sclerotia germinated over a temperature range of 16-30°C with the optimum at 28-30°C.

pH

The minimum, optimum and maximum pH for growth of ShB fungus was 2.5, 5.5 and 7.8 respectively (Endo, 1935). Misawa (1965) reported that utilization of different carbon sources in the medium was affected by its pH. Sherwood (1970) reported high adaptability to pH extremes, as fairly good growth was obtained even when pH drops as low as 2.5 in the media containing ammoniacal nitrogen such as NH_4NO_3 and $(NH_4)_2SO_4$ as well as $NaNO_3$ and glutamic acid, where the pH rises up to 8.7.

Carbon utilization

While studying the nutritional physiology of ShB fungus, among the several sugars tested Manian (1982) reported that glucose, mannose, sucrose, maltose, cellobiose, starch and cellulose promoted good growth, least by fructose, while sorbose totally inhibited. Better growth was seen in glucose than in maltose or in starch because of the simple structure of glucose compared to maltose or starch which has to be cleaved first and then assimilated. Despite being a successful saprophyte, surprisingly *R. solani* isolates poorly utilised the soluble cellulose (CMC). Misawa (1965) reported that monosaccharides promoted better growth than disaccharides but a few polysaccharides gave the best growth. Yoshimura (1955a) studied the ability to decompose cellulose but not inulin. Manian (1982) observed that five differentially virulent isolates R1 to R5 obtained from different locations (Table 3) designated as per the virulence ranking (Zuber and Manibhushanrao, 1982) (more details are given in Chapter 6, on Pathogenicity and Fig. 6) differed significantly in their response to different forms of carbohydrates and their ability in the formation of sclerotia. Isolate R3 with intermediate virulence showed maximum mycelial dry weight and sclerotial initiation followed by highly virulent R5 and least virulent R1. However, the growth measured as mycelial dry weight could not be correlated with linear growth rate.

-Table 3: List of *R. solani* isolates with their respective source and virulence (Zuber and Manibhushanrao, 1982).

Isolates	Source	Virulence*
R_1	University Botany Laboratory Culture Collection, (now CAS in Botany) University of Madras, Madras	4.39
R_2	Central Rice Research Institute, Cuttack, Orissa	5.18
R_3	All India Coordinated Rice Improvement Project, (now DRR) Hyderabad, Andhra Pradesh	6.75
R_4+	Paddy Experiment Station, Thirurkuppam, Tamil Nadu	9.23
R_5	University of Agricultural Sciences, Bangalore, Karnataka	10.62

* On Ponni susceptible cultivar virulence was expressed as disease severity index (DSI) = mean lesion length in centimeters × average number of lesions per inoculated tiller.
\+ Isolated by the authors from diseased rice plants.

Nitrogen utilization

Manian (1982) observed that peptone and glutamic acid are superior to inorganic sources viz. $NaNO_2$, NH_4NO_3 and $(NH_4)_2SO_4$. On the contrary, inorganic NH_4NO_3, $NaNO_3$ are the best sources for sclerotial initiation. Fungi unable to synthesize nitrite reductase are very sensitive to $NaNO_2$. However, the differentially virulent isolates grew well in the $NaNO_2$ medium and perhaps such adaptation might be the reason to include this salt in the selective medium for the isolation of ShB fungus from soil (Ko and Hora, 1971). Misawa (1965) reported that organic nitrogen source promoted better growth than inorganic sources of nitrogen and no appreciable difference was found in utilizing NH_4NO_3 and KNO_3. Arginine, urea, threonine, glycine and $(NH_4)_2SO_4$ were the best sources of nitrogen for mycelial growth (Santos, 1970).

Inoue and Uchino (1963) reported that $(NH_4)_2SO_4$ and peptone in the medium inhibited sclerotial formation. Santos (1970) reported that the size and number of sclerotia formed on agar plates were affected by the type of carbon and nitrogen sources present in the medium. Proline could induce many (average 37 per petridish) very large size sclerotia (3.5 x 3.1 mm) per petridish. A few (3-8) large (2.5 x 2.2 mm) sclerotia were produced on urea, isoleucine, threonine, arginine, leucine and valine incorporated media. Abundant (162-244) moderate sized (2.3 x 3.4 mm) sclerotia were produced on $NaNO_3$, serine, glutamine, alanine and aspartic acid supplemented media. Few or no sclerotia were produced in tryptophan, histidine and lysine amended media. Manian (1982) reported that no definite correlation could be made between virulence of the isolates and the types of C/N source utilized.

Toxins

Newton and Mayers (1935) were the first to detect a 'toxic principle' in the culture filtrates (CF) of *R. solani*. Although Chen (1958), Aoki et al. (1963), Nishimura and Sasakii (1963) and Nour El Dein and Sharkas (1965) reported the presence of non-enzymatic, low molecular weight phytotoxic metabolites *in vitro* namely, phenyl acetic acid (PAA), and its *para, meta*-hydroxy derivatives as well as *in vivo* particularly *p*-HPAA and *m*-HPAA (Kohmoto et al., 1973). Ramalingam (1983a) could not detect *p*-HPAA *in vitro* except *m*-HPAA in repeated tests and little is known about their mechanism(s) in causing disease syndrome. A comprehensive account on phytotoxic metabolites of *R solani* has been provided by Manibhushanrao et al. (1981).

Manibhushanrao and Ramalingam (1992) obtained a positive correlation between toxin production and the relative virulence of the ShB fungus isolates (R1 to R5). A number of toxic metabolites were detected from the CF of the pathogen grown in potato dextrose broth (PDB) amended with 0.5% peptone, whereas *p*-HPAA was the only toxic metabolite detected in the Czepax Dox (CD) broth. The production of phytotoxin(s) by ShB fungus was greatly influenced by the substrate on which it was grown. The factors also influenced *in vitro* toxin production.

Toxin production

A number of factors namely time, nature of the culture filtrate, pH of the medium, temperature, aeration (stationary and shake cultures) influenced the optimal production of the toxin *p*-HPAA (Ramalingam, 1983a). No positive correlation was obtained between the growth of the fungus and *p*-HPAA production. However, toxin production increased irrespective of aeration over a period of time.

In shake culture, accumulation of the toxic compound was observed after only 5 days. The optimum pH and temperature were 6.5 and 25°C respectively for maximum toxin production. Although the growth of the fungus was atleast two to three-fold more in shake culture, no appreciable difference was observed with reference to the production of toxin between stationary and shake cultures (Fig. 2a). Growth as well as the toxin production were best in the pH range 6.0-7.0 with optimum at 6.5 and 25°C (Fig. 2b). $NaNO_3$ was the best source for maximum toxin production. Though phenylalanine promoted appreciable replication of DNA in the fungus, it totally inhibited the production of the toxin. Both glucose and sucrose were the best carbon sources for good growth as well as for the maximum toxin production.

The failure to produce *p*-HPAA by the least virulent isolate (R1) either in stationary and shake culture (Fig. 3a, b) or in rice soil medium, further substantiates the role of a possible genetic factor for the production of toxin. However, the quantitative correlation of the virulence of the isolates with toxin production is not easily interpreted since toxin production *in vitro* is influenced by the composition of the medium and the physical environment (Ramalingam, 1983b).

Permeability and respiratory changes

Ramalingam (1983a) reported that all the four toxic metabolites tested, triggered heavy leakage of electrolytes from detached as well as coleoptiles of the susceptible cv. TKM 9 at higher concentration (1000 ppm) in the initial period (4 h) of the toxin treated rice tissue. However, at low concentration (100 ppm) a sharp increase in the leakage was observed with *m*-HPAA and *o*-HPAA at 8 h treatment followed by a sharp decline at 16 h.

The host tissue respiration was also stimulated markedly at 100 ppm concentration while inhibition of O_2 uptake was noticed at 1000 ppm. Addition of 2-4 dinitrophenol

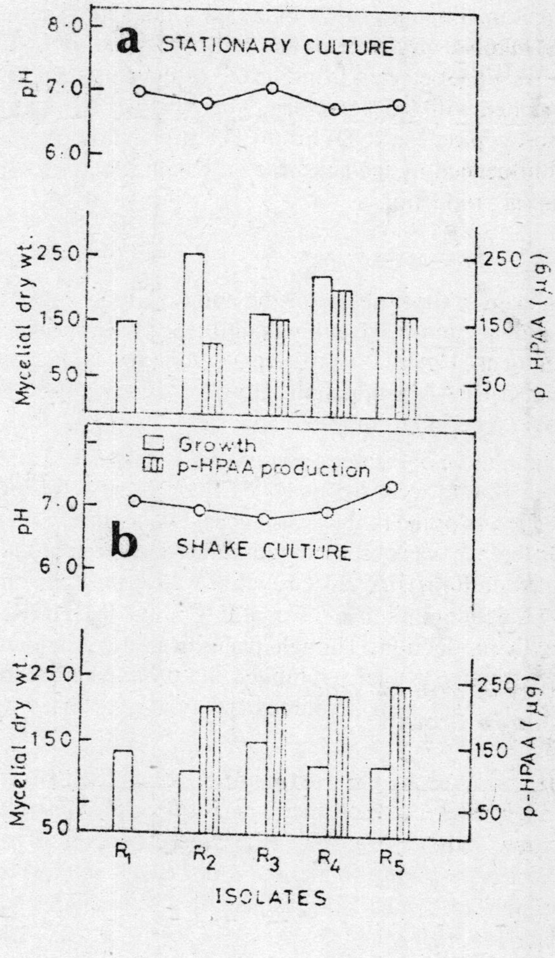

Fig. 2ab : Changes in the growth and toxin production by *R. solani* (a) stationary and shake culture, (b) various pH and temperature (Manibhushanrao and Ramalingam, 1993).

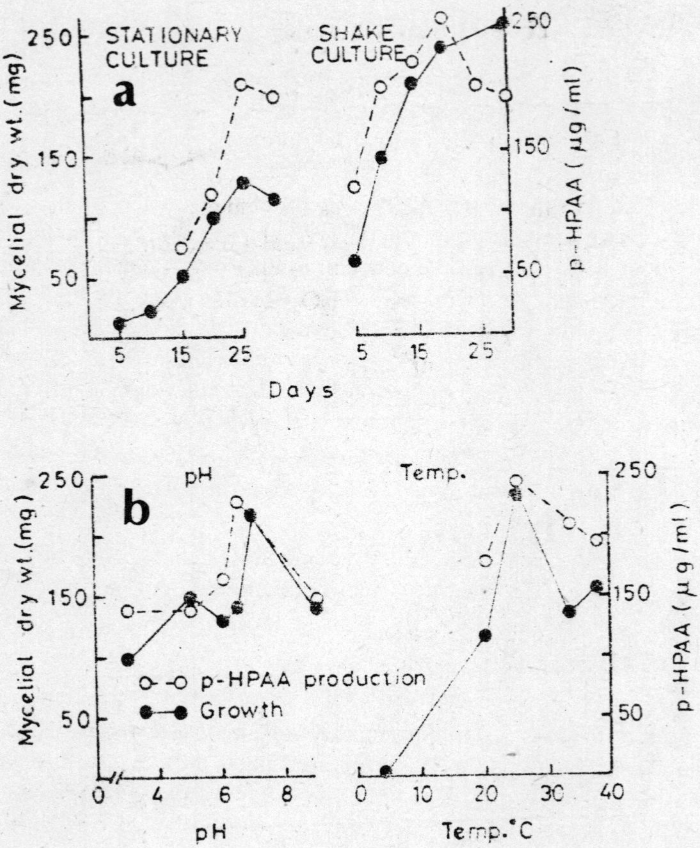

Fig. 3ab : Growth response and toxin production by *R. solani* isolate grown in CD broth (a) in stationary and (b) shake culture (Manibhushanrao and Ramalingam, 1993).

(DNP), a protein inhibitor to the toxin treated rice tissue did not bring out any additive effect. The lesser response of toxin treated tissue as well as *R. solani* inoculated plant tissue towards DNP appears to be due to the uncoupling of oxidative phosphorylation which probably explains an increase in the infected tissue respiration. The changes in permeability and tissue respiration in the presence of the toxic metabolites indicate the initial response compared to that of *R. solani* infection on rice.

Key enzymes of respiration and phenol metabolism

Ramalingam (1983a) further reported that PAA and its derivatives stimulated ascorbic acid oxidase (AO), catalase (CA) and peroxidase (PO) activities at 100 ppm concentration. However, at 1000 ppm concentration a total inhibition of the activities of these enzymes were noticed with few exceptions, *p*-HPAA at 1000 ppm, *o* and *m*-HPAA at both the concentrations. On the other hand, there was no change in the treated phenylalanine ammonia lyase (PAL), tyrosine ammonia lyase (TAL) activities following the treatments of these toxic metabolities at both the concentrations used. It appears that increased CA activity might have helped to decrease the H_2O_2 level in the cells thereby limiting the peroxidative reactions catalysed by PO. It is also evident that enzymes CA and AO may play a role in O_2 consumption. Reduction in PO activity might account for inhibition of O_2 uptake following toxin treatment. Further studies on these respiratory enzymes in relation to their substrates may reveal more on the respiratory metabolism of rice plants.

The stimulation of PAL and TAL activities may be related to the phenyl propionoid biosynthesis by the toxic metabolites at low concentration and could be compared with the phytoalexin elicitors.

These results further suggest that the PAA and its derivatives may induce, enhance or block the synthesis of specific enzymes or may regulate enzymatic activities through action or repression and warrants further study.

Biosynthesis of aromatic compounds

The production of toxic metabolites mainly depend on the host-parasite combination. Manibhushanrao and Ramalingam (1993) reported that in the *R. solani* infected susceptible TKM 9 and resistant IR 20 rice leaf tissue, both phenylalanine and tyrosine were not detected. Failure to detect these two aromatic compounds in the host might be attributed to their quick conversion to the other hydroxy isomers. Further, *in vitro* and *in vivo* studies on biosynthesis of these aromatic amino acids by ShB fungus were carried out with replacement culture techniques as well as by incorporating labelled precursors. In the replacement culture studies with ^{14}C-phenylalanine, the intermediary compounds identified were *p*-hydroxybenzaldehyde, *m*-HPAA, protocatechuic acid, *p*-coumaric acid, PAA, cinnamic acid and *o*-HPAA while with ^{14}C-tyrosine, *p*-HPAA, *p*-coumaric acid and protocatechuic acid.

The identification of *m*-HPAA, *o*-HPAA as intermediate compounds of phenylalanine in addition to other compounds does not rule out the operation of yet another pathway. The results of replacement culture studies indicate that, ShB fungus employs both transamination and non-oxidative deamination to deaminate phenylalanine, hydroxylation of PAA to its *meta*-position which appears to be characteristic of *R. solani* isolates and may play an important role in pathogenesis (Kohmoto and Nishimura, 1974). In addition, it helps to identify the particular pathway leading to *m*-HPAA production which might be involved in pathogenesis. Tyrosine degradation is probably through the formation of *p*-coumaric acid and results in accumulation of protocatechuic acid.

Phytotoxicity of the metabolites

All the intermediate compounds of the phenyl-pyruvate pathway and *p*-HPAA were phytotoxic by inhibiting the root and shoot elongation of rice seedlings and also induced typical lesions on excised leaves of susceptible cv. TKM 9. However, at 10 ppm and below, these compounds did not produce any visible symptoms on both TKM 9 and IR 20 but inhibited root elongation (50%) and slightly stimulated the shoot elongation, like any other growth promoting hormone. Toxins of ShB fungus induce lesions comparable to the lesions produced by the pathogen produced (Plate 5 ABC) and appear to play a role in disease development (Manibhushanrao and Ramalingam, 1992). Nevertheless, apart from action of toxins in the induction of symptoms in the host, the role of polysaccharides cannot be ignored. In the pathogens' armoury whether these toxins act as a primary toxic weapon along with pectic polysaccharides on the host tissue prior to hyphal penetration or not, need further ultrastructural and histochemical studies.

Mycelial proteins, enzymes and amino acids

Zuber and Manibhushanrao (1982) reported the polyacrylamide gel electrophoretic patterns of proteins and various enzymes from mycelial and extracellular extracts of five differentially virulent ShB fungus isolates. There were marked variations observed in the protein (Fig. 4) and isoenzyme patterns of polyphenol (DOPA) oxidase, peroxidase, acetyl esterase and acid and alkaline phosphatases from mycelial and extracellular extracts. There was no correlation between the enzyme patterns and the virulence of various isolates. Less virulent isolates showed mycelial and extracellular peroxidase bands and only the least virulent isolate gave mycelial alkaline phosphatase bands.

Two bands with Ef 0.6 and 0.8 with increasing intensity from R1 to R5 can positively be correlated to increasing virulence of the respective isolates while band at Ef 0.42 appears to be a product of gene for avirulence. These protein bands might serve as biochemical markers to identify the different isolates of ShB fungus. Acetyl esterase isoenzymes were detected in mycelial and extracellular sources of these isolates and none of these isolates showed an isoenzyme band. Contrary to the marked variation of ShB fungus isolates from rice, Matsuyama et al. (1978) showed similarity in isoenzyme patterns of extracellular acetyl esterase in ShB fungus isolates from rice as well as other cereals.

Acid phosphatases appear to be restricted to fungal mycelia alone but not excreted into culture medium. Isoenzymes of alkaline phosphatases appeared mainly in fungal culture-filtrates, and they were also visualized in the mycelium of R1 as well. This can be possibly be attributed to the fact that a single species sometimes involves several physiologically different isolates, despite their morphological similarities (Matsuyama and Wakimoto 1977). Instead the problems concerning variability within a species can be the result of analysis of only a few isolates of a single species or of a limited number of enzymes studied. Ramalingam (1984a) studied the effect of phytotoxic metabolites of the fungus on

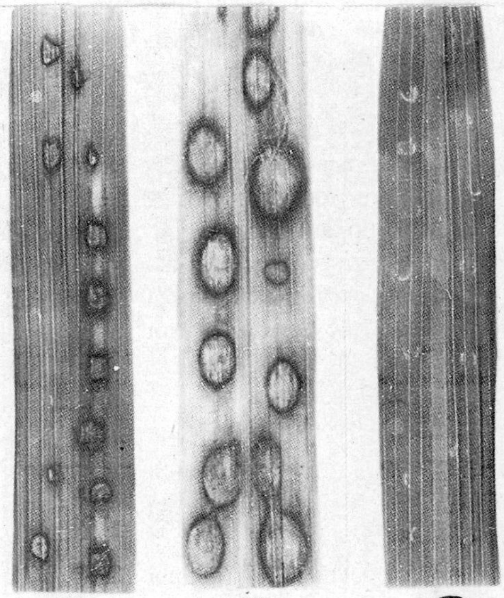

Plate 5 AB : Induction of typical necrotic lesions on the susceptible rice cv. TKM 9 leaves. (A) TLC eluted compounds, (B) authentic *p*-HPAA (1000 ppm), (C) water control (Manibhushanrao and Ramalingam, 1993).

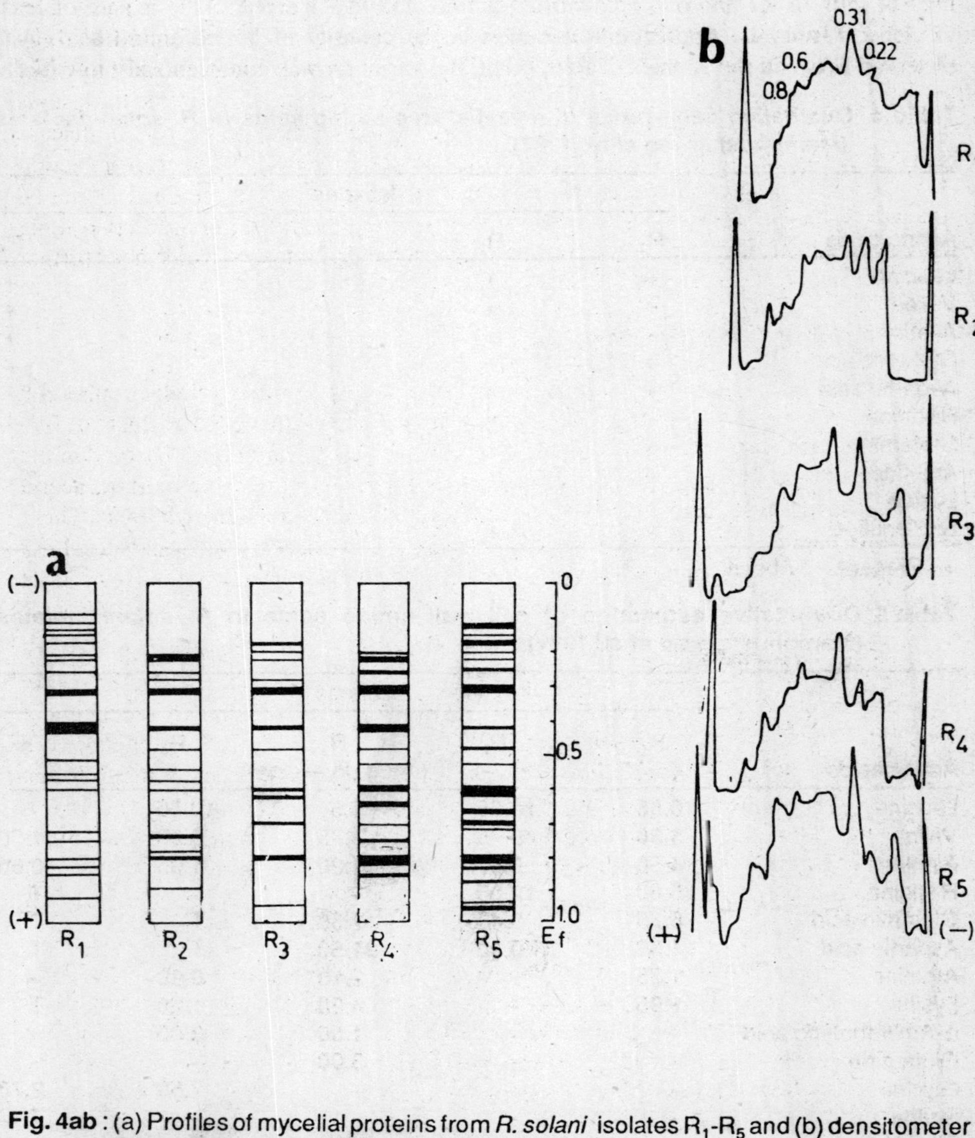

Fig. 4ab : (a) Profiles of mycelial proteins from *R. solani* isolates R_1-R_5 and (b) densitometer scanning (Zuber and Manibhushanrao, 1982).

isoenzyme patterns of peroxidase. He also reported the production of polygalacturonase and cellulolytic enzymes by ShB fungus (Ramalingam, 1983b, 1984b).

Manibhushanrao et al. (1987) reported the free and bound mycelial amino acids of the five differentially virulent isolates (R1 to R5) ranged between 5-8 and 6-10 respectively. Of the mycelial free amino acids identified, only alanine, glutamic acid and aspartic acid were common to all the isolates and no specific trend was observed with either the presence or absence of other amino acids (leucine, valine, histidine, glutamine and arginine). However, lysine was present only in high virulent isolates, R4 and R5; glutamine was present only in R1 and one unidentified amino acid in R1 and R2, the isolate of least virulence (Table 4). Appreciable variation in the quantity of bound amino acids was observed between the isolates (Table 5) but, the variation was conspicuously low in the

Table 4: Qualitative occurrence of mycelial free amino acids in *R. solani* isolates (Manibhushanrao et al. 1987).

Amino acids	Isolates				
	R_1	R_2	R_3	R_4	R_5
Leucine	+	+	+	-	+
Valine	+	+	+	-	+
Alanine	+	+	+	+	+
Glutamic acid	+	+	+	+	+
Aspartic acid	+	+	+	+	+
Histidine	-	+	+	+	+
Glutamine	+	-	-	-	-
Arginine	+	-	-	-	+
Lysine	-	-	-	+	+
Unidentified	+	+	-	-	-

+ : Present; - : Absent

Table 5: Quantitative estimation of cell wall amino acids in *R. solani* isolates (Manibhushanrao et al. 1987).

Amino acids	Isolates				
	R_1	R_2	R_3 (mg/g dry weight)	R_4	R_5
Leucine	10.65	10.05	7.5	10.65	-
Valine	3.60	3.75	3.75	3.60	1.50
Alanine	4.50	2.50	1.20	1.05	0.60
Histidine	25.80	12.00	-	-	T
Glutamic acid	6.75	0.45	2.55	T	T
Aspartic acid	1.50	0.30	31.50	T	T
Arginine	1.35	-	2.10	0.90	-
Lysine	1.95	-	4.20	1.20	T
α-Aminobutyric acid	-	-	1.50	3.00	-
Glutamine	-	-	3.00	-	-
Glycine	-	-	-	7.50	2.70
Serine	-	-	-	-	T
Phenylalanine	-	-	-	T	-

T : Traces; - : Absent

virulent isolate R5. The results indicate that differentially virulent *R. solani* isolates from rice exhibited greater qualitative and quantitative differences than similarities in their amino acid composition. However, Reddy and Rao (1975) reported specific correlation with the greater production of amino acids by the virulent *R. solani* isolates from groundnut.

Sclerotial growth and maturation

The physiology of sclerotial growth and maturation are complex processes. The sclerotia has to pass through three phases before maturation. Initially a rapid RNA synthesis occurs which is followed by a decline in poly A (+) - RNA and a decrease in the permeability of sclerotia thereby indicating that the first 2 h represent a period of reorientation of cellular metabolism followed by a period of general synthesis. This process lasts for 30 h which is accompanied by further decrease in permeability. Subsequently, an increase in protease and ribonuclease activity and rapid build up of ribosomes occur. Then there was a decline in the protease activity where the other enzyme activities maintain and the sclerotia began to turn brown. Maturation of sclerotia last for 15-ds which then begin to float in water.

Hemmi and Endo (1928) devised a staining method to test the viability of the sclerotia where the dead ones take up the stain while the live ones did not. Further, Hemmi and Endo (1931) found that light is essential for the formation of sclerotia. Most of the mature sclerotia float on water (Hashiba et al., 1972; Hashiba, 1982a; Hashiba and Mogi, 1975; Lee, 1980) and there was a close correlation between density of sclerotia in a field and the severity of the ShB (Hori and Anraku, 1971; Yamaguchi et al., 1971; Hashiba, 1984a).

The pathogen was most frequently isolated from overwintered sclerotia with about 67.7% germination at two months after collection, 42-78% at post-harvest paddy fields while in the uncultivated paddy fields about 60% in the 1st year and 40% in the second year (Hashiba and Mogi, 1973).

Temperature is known to influence the survival of sclerotia to a considerable extent (Hashiba and Yamaguchi, 1971; Nishikado and Hirata, 1937; Prabath et al., 1974) and the role of mycoparasites and mycophagus animals in causing destruction of sclerotia have been well established (Coley-Smith and Cooke, 1971). Information on the use of such microorganisms to control ShB fungus is meagre. Manian and Manibhushanrao (1990) reported the influence of some factors on the survival of ShB fungus in soil. Sclerotia remained viable for atleast 21 months under moist conditions in a typical rice field soil at two controlled temperatures of 20 and 30 ± 2°C, but survival was adversely affected under dry soil conditions at 30 ± 2°C. Further, the survival of sclerotia in the wet soil at 30°C is comparable to that at 20°C in wet and dry soils (Fig. 5). Additionally, the synergistic effects of two unfavourable conditions, dryness and high temperature leading to early loss of viability of sclerotia, is also apparent in the study. Sclerotial viability was not affected by soil microorganisms as there was no significant difference in their survival under sterile or non-sterile conditions. However, Roy (1985a) reported the poor competitive saprophytic activity of the sclerotia in the presence of soil inhabiting microorganisms.

Density vs viability

Information on viability of sclerotia of ShB fungus on rice plants, culture media as well as in the soil is available (Tu et al., 1979; Hashiba and Yamada, 1981; Manian and

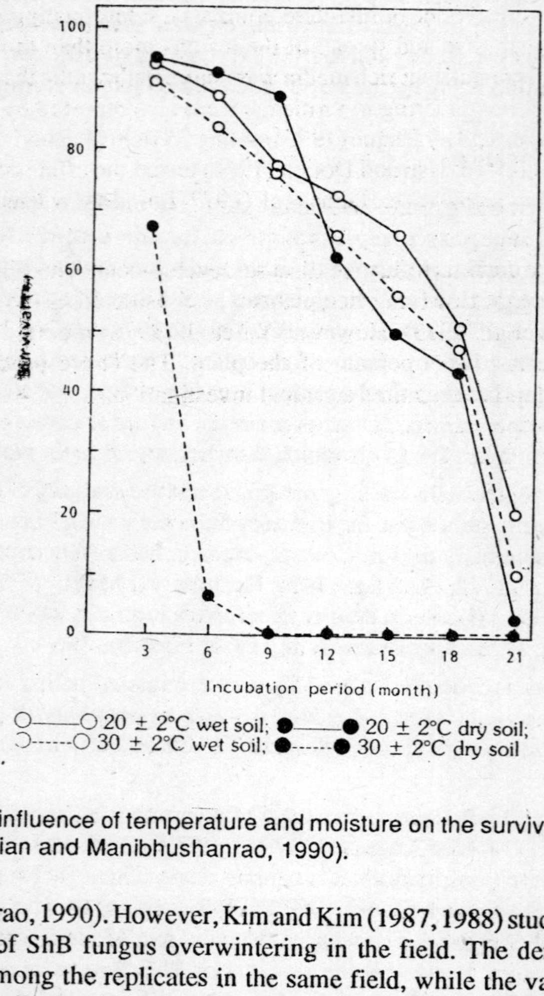

Fig. 5 : The influence of temperature and moisture on the survival of sclerotia of *R. solani* (Manian and Manibhushanrao, 1990).

Manibhushanrao, 1990). However, Kim and Kim (1987, 1988) studied the sclerotial density and viability of ShB fungus overwintering in the field. The density of sclerotia varied appreciably among the replicates in the same field, while the variations in their density between the soil samples collected within the fields tend to decrease but the density increased.

Mgonja (1983) reported the relationship of rice ShB incidence and sclerotial density and distribution of ShB fungus in Arkansas, USA. Murata and Nagano (1985) obtained a relationship between density of the sclerotia and ShB injury with seedling density in a direct sowing system in Japan.

Variability

The fungus isolates are highly variable in their aggressiveness (Amaral et al., 1979; Zuber and Manibhushanrao, 1982; Dath, 1984). Santos (1970) found that C and N sources have great influence on pathogenicity. Isolates grown on media containing arginine, urea, threonine and glycine were more virulent than those cultured on media containing serine, isoleucine, tryptophan, tyrosine, lysine or histidine. Similarly, sclerotia dipped in nutrient solution increased the number, size and spread of the lesions more than those dipped in water. Again, the sclerotia from nutrient rich media were more pathogenic than those from poor media (Yu, 1975). Isolates differing in virulence were also reported by Tsai (1973) among 40 single spore cultures and by Haque (1975) among 25 field isolates but no distinct differential reactions were noticed. Hsu and Dough (1964) tested the effect of Hg, Na, Ni, Sn, Pb, Fe and Al salts on *R. solani* growth. Haque et al. (1977) found a few sensitive isolates to high concentration of CO_2.

Host nutrition is also known to influence disease development. The higher nutrient content of upper leaf sheaths of flowering rice plants was considered as the main reason for host susceptibility (Ou et al., 1973). However, Yu et al. (1976) reported that disease development was not affected by N content of the plant. The entire problem of host nutrition and disease development requires a critical investigation.

Pathogenicity

Quantification of rice ShB disease is quite essential for the evaluation of the yield losses accurately. Many related terms such as pathogenicity and virulence, disease intensity and disease severity are used in the assessment of the plant disease by many pathologists. Detailed information on such terminology used in relation to rice blast disease is discussed (Manibhushanrao, 1994).

The fungus is highly variable in its virulence and at times it is associated with hyphal growth. Isolates with less aerial mycelium were highly virulent (Tu, 1967) while the reverse was also reported (Akai et al. 1960). No correlation could be obtained with the ability to produce pigment in the culture media and pathogenicity (Tu, 1967).

Pathogenicity of isolates was correlated with phosphatidase production. Highest phosphatidase activity is associated with initial states of disease development (Kuan and Kuo, 1974). Increased pathogenicity in terms of size and number of lesions was observed by spraying the rice plants with 2, 4-D regardless of the time after application. In addition, a decoction of rice plants sprayed with 2, 4-D stimulated the growth of *C. sasakii* (Kurodani et al., 1959). The mechanism of increased susceptibility in rice plants to 2, 4-D is not known.

Chien and Chung (1963) tested 300 isolates from Taiwan on 16 cvs. and based on the degree of pathogenicity (number of leaves infected) they classified into 7 cultural types and 6 physiological races and found a relationship between cultural characters and races. These observations are based on visual rating. Saikia and Roy (1975a) reported the differential pathogenicity of *R. solani* on a few host plants.

Zuber and Manibhushanrao (1982) collected several ShB fungus isolates of rice from south India and tested their pathogenicity on susceptible rice cvs. Ponni and TKM 9 as well as on moderately resistant IR 20. Virulence, the disease-inducing capacity or quantified differences in disease severity is generally expressed as disease severity index (DSI) i.e. mean lesion length in cm x average number of lesions per incculated tiller. The relative virulence i.e. from 4.93 to 10.62 increased from R1 to R5 (Table 3). Quantitative studies revealed that all the isolates were more aggressive on Ponni than on IR 20 (Fig. 6) and the trend of virulence ranking was almost the same thereby suggesting the lack of any physiological specialization. However, critical work on more differentially virulent isolates on several cvs. should be done for any positive conclusion. Nandi and Chakrabarthi (1984) reported differing virulence in the isolates of *C. sasakii* to rice cvs. and the type of resistance offered by host plants.

Narayanaswamy and Rao (1984) reported a new fungus *Waitea circinata* causing ShB of rice, similar to that of symptoms produced by the rice ShB fungus. Nonaka (1984) studied the incitant *R. solani* (*sasakii* type) and its relation to ShB-like disease of rice plant.

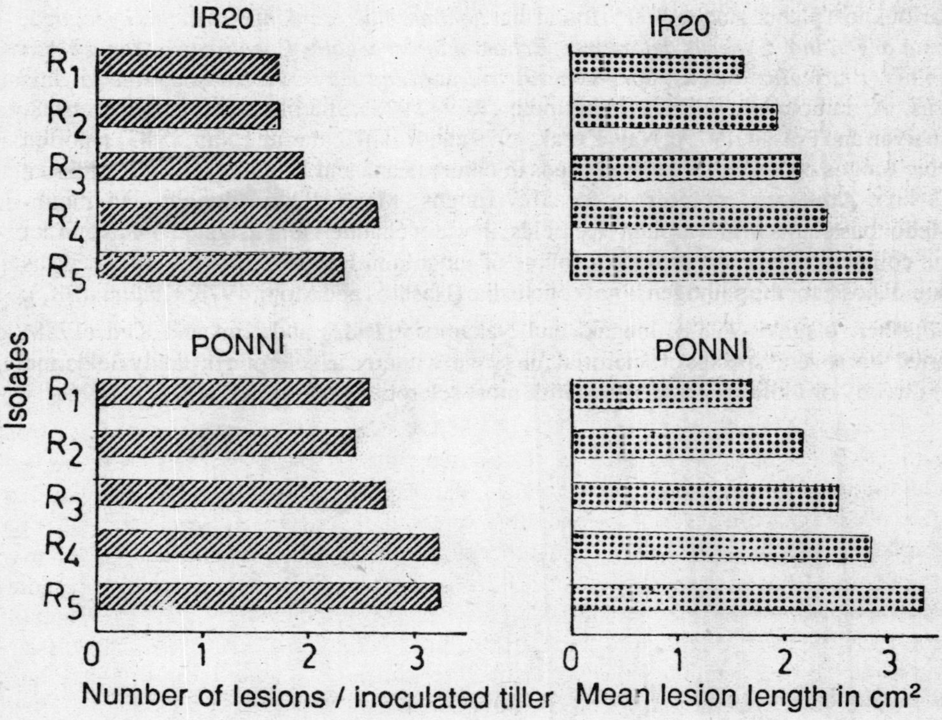

Fig. 6 : Virulence ranking of *R. solani* isolates (R_1-R_5) on two rice cvs. IR 50 and Ponni (Manian, 1982).

Host Range

The ShB fungus infects a number of plants, and similarly *R. solani* isolates from other crops are able to incite the disease on rice (Yokogi, 1927; Tervet, 1937; Ryker, 1939 and Sato and Shoji, 1957). It was also reported to infect a wide range of plants comprising 188 species in 32 families (Kozaka, 1965). Hori (unpublished) tested 18 species and 817 cvs. of rice and reported that none of them was physiologically resistant to ShB. Tasi (1970) studied the host range on 20 species of weeds belonging to 11 families of which Gramineae and Cyperaceae were dominant. Saikia and Roy (1975b) reported that it is pathogenic to 60 different hosts from 19 families. Chin and Lim (1975) studied the host range of *R. solani* on various host plants. Haque (1975) found that *Abelmoschus esculentus, Arachis hypogaea, Celosia argentina, Cyperus deformans, Echinochloa crusgalli, E. colonum, Monochoria vaginalis, Phaseolus aureus, Sorghum vulgare* and *Zea mays* were susceptible to ShB fungus. A number of workers from India (Roy, 1973; Sharma and Mukherji, 1978; Kannaiyan and Prasad, 1979c; Nayak et al., 1979 and Gokulapalan and Nair, 1983) reported that the fungus occurred on several weeds in nature. Lana and Khatua (1984) identified a grass host *Axonopus campressus* for ShB fungus. Many dicotyledonous and monocotyledonous plants in and around rice fields, in water channels and irrigated ponds can act as the collateral hosts and provide a source of inoculum. However, the role of weeds as collateral hosts for the pathogen is not conclusive (Hashiba and Mogi, 1973; Khatua, 1982).

Further, Inagaki (1983), Inagaki and Nakamoto (1982) and Kim and Kim (1988) reported that several species of sclerotial fungi overwintered as sclerotia in paddy fields and were directly or indirectly associated with most sclerotial diseases.

Environmental Conditions

Kozaka (1965) reported that ShB fungus incite disease on rice at a temperature range 23-35°C (optimum 30-32°C) and RH (96-97%). Earlier, Endo (1930), Hemmi and Endo (1933) recorded that at 32°C infection took place in 18 h and at 28°C in 24 h with continuous wetting.

It is well established by many workers that the incidence of the disease is most severe under conditions of high humidity and temperature. The temperature within the rice crop varies with the prevailing air temperature while close planting and heavy application of N-fertilizers lead to thick growth at the water front and tend to increase the relative humidity. Kozaka (1961) reported two distinct phases of disease development, the vertical i.e. upward spreading of infection to upper leaf sheaths and horizontal infection spreading i.e. to the neighbouring plants and fields. Both these developments occur rapidly under favourable conditions while the vertical spread generally occurs only after heading stage. If the conditions are conducive then sclerotia will be seen on the upper leaf blades, otherwise they are restricted to the lower sheaths. Endo (1935) and Yoshimura (1955b) reported that direct sunlight inhibits infection while shade promotes it.

High N content in the soils seems to increase the disease severity (Kozaka, 1961, 1965 and Loo et al., 1963) and the susceptibility is directly correlated with the increased N level than that of carbon (sugar) (Kozaka, 1961). Amount of phosphate also has a direct correlation with the disease incidence (Inoue and Uchino, 1963) but addition of salt content (0.01 - 1% NaCl) reduced the infection (Endo, 1933)

The intensity of primary infection rests mostly on the number of sclerotia which are in contact with the plant at the water surface. Subsequent disease spread is governed by the favourable environmental conditions as well as susceptibility of the host plant. The rice seedlings at dapong nurseries were proved to be disease prone because of their compact growth. It has been reported that close planting of rice plants also become susceptible as they grow older (Kozaka, 1961) perhaps due to high humid condition in both the cases.

Disease Cycle

Sclerotia are the main source of inoculum. The fungus overwinters as sclerotia or as mycelium in soil (Endo, 1931). The sclerotia lose their viability in shorter periods as the soil temperature increases (Nisikado and Hirita, 1937, Nishikado et al., 1938; Roy, 1976). The sclerotia are easily detachable from the host and, as they fully mature they fall, sink, and soon float on the water surface (Malaguti, 1951). They remain viable in soil for several months over a wide range of temperature and moisture (Palo, 1926; Endo, 1931 and Park and Bertus, 1932).

Agricultural operations such as ploughing, transplanting, weeding, and levelling help the sclerotia to surface and make initial contact with the host. The germination of sclerotia is also important. Hori and Anraku (1971) reported that majority of the sclerotia (60-70%) found less than one cm beneath the soil surface were less viable while those found in the lower layer seems to be more viable with 30-50% germination. Hashiba et al. (1972) showed a decrease in sclerotial germination with raise in temperature. Yamaguchi et al. (1971) reported that in Japan a mean of 57 sclerotia were produced per hill in a plot of heavily fertilized and severely infected fields. They estimated that 10% of the sclerotia formed on plants floated off on to the water after field operations. Similar estimates were made by Tu et al. (1979) in Taiwan. The specific gravity of sclerotia from the field was less (0.8-0.9) while those from the culture was more (1.1-1.3). The sclerotia when cut into 2, 4, 8 or even 16 pieces still remained capable of infecting the plants. However, a positive correlation was observed between the size of these pieces and the lesion length (Chien et al., 1969). The sclerotial distribution in the field were estimated by scooping and suction (Yamaguchi et al., 1971) as well as other methods (Hori and Anraku, 1971). Of the two methods the former one was most efficient and simple than the latter. Sieve-floating method developed by Ui et al. (1976) seems to be more accurate in determining the sclerotial populations in soil besides, they retained their infectivity.

The disease spreads primarily through soil by sclerotia but also seedborne and rarely airborne. Basidiospores occasionally incite infection. Although Saksena (1972) reported infection by basidiospores, Fukatsu et al. (1960) found it difficult while Ou (1985) failed to infect the leaves and sheaths of rice plants.

There are reports on infection of the rice ShB fungus on panicles before grain filling (Marchetti, 1983; Ou, 1985) and on seeds (Kannaiyan and Prasad, 1978). The source of seed-borne infection might be due to direct contact with the infested soil (Roy, 1989). The key events of the ShB disease cycle is given in Fig. 7 (Manibhushanrao, 1990).

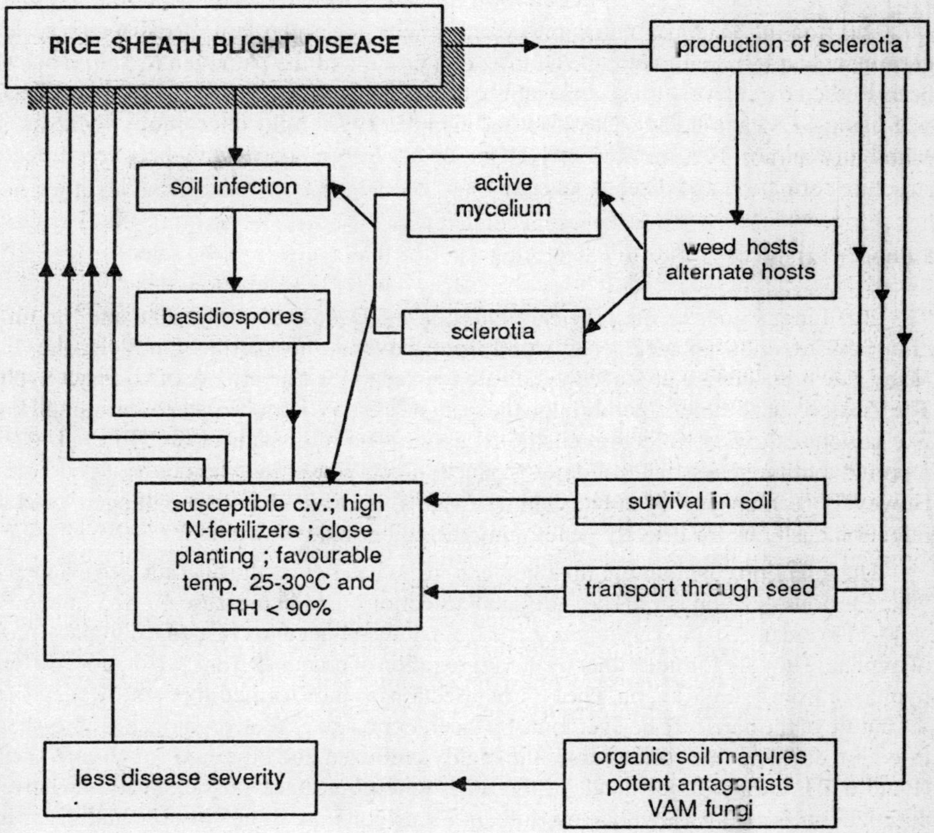

Fig. 7 : Key events of the rice ShB disease cycle (Manibhushanrao, 1990).

Histopathology

Host penetration

The floating sclerotia come into contact with the rice plant at the water surface, germinate and initiate infection. The infection process of the pathogen *R. solani* has also been studied extensively on several other hosts (Dodman and Flentje, 1970) as well as on rice using SEM and TEM (Hashioka and Okuda, 1971) light microscopy (Manian and Manibhushanrao, 1982a). Recently, Kim et al. (1990) correlated between infection structure formation and disease severity.

Pathogen on the host prior to penetration

ShB fungus with its primary, straight, non-infectious runner hyphae and the multi-branched, infectious lobate hyphae usually spread over the host surface parallel to the veins. Many side branches are produced which are twisted to form a network of compact hyphae. The lobate hyphae either enter through the stomata as such or proliferate to form small knot-like structures called 'infection cushions' about 40 h after incubation. Ou et al. (1973) reported that the rice isolates did not produce infection cushions before host penetration. However, Manian and Manibhushanrao (1982a) showed that rice pathogen produces infection cushions for directly penetrating the epidermal cell wall.

The infection cushions clump together to form patches on the host surface (Plate 6 A), where lesions develop later. The individual infection cushion measure 40-325 (mean 120) X 30-115 (mean 55) µm, varying in shape from dome-shaped to crust-like compact masses of hyphae. They are formed either by the aggregation of branches from two or more different hyphae or from a single hypha. The type of infection cushion formed appears to depend upon the nature of the barrier to be overcome by the infection-peg. Well organized, dome-shaped infection cushions are formed over the highly cutinized and silicified rough leaf surface (Plate 6 B). On the other hand, in the inner smooth surface of the sheaths, where the cutinization is comparatively low, infection cushions are small in size and crust-like, lacking definite organization of cells. Their structure is relatively simple unlike the upper horizontal and basal vertical arrangement of cells as seen in the dome-shaped cushions. These are also produced on artificial surfaces such as a glass surface (Plate 6 C) or a collodion membrane (Plate 6 D) (Manian and Manibhushanrao, 1982).

The factors which govern the formation of infection cushions are complex. Marshall and Rush (1980 a and b) suggested a relationship between infection cushion formation and host exudates as well as varying levels of resistance of rice cvs. to ShB showing various amounts of wax deposits. Though ShB fungus is known to produce cutinolytic enzymes (Parmeter, 1970), cuticle appears to be more resistant to enzymatic degradation than the other cell wall constituents, cellulose and pectin. The production of these enzymes by the

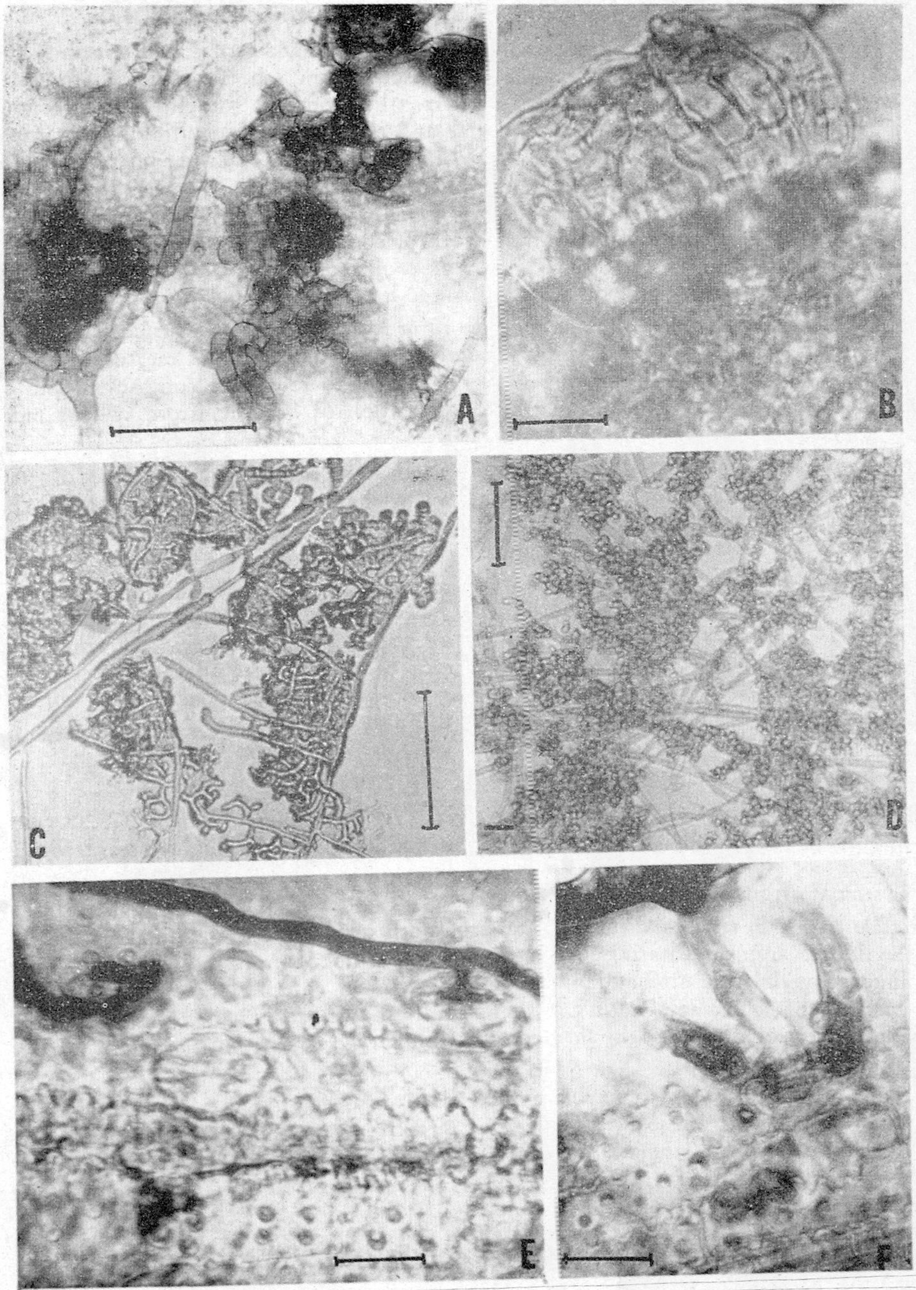

Plate 6 ABCDEF : (A) lobate mycelium of *R. solani* producing infection cushions, (B) vertical section through dome-shaped infection cushion showing upper horizontal and lower vertical layers of thick walled cells, (C) and (D) infection cushions formed by contact stimulus over the glass and collodion membrane respectively, (E) and (F) showing the entry through stomatal openings. Note : 'stomatal tropism' - attraction of one or more lobate hyphae with or without producing appressoria-like structures at the point of penetration through stomata. Bar : A and D-100μm; C-200μm; B, E and F-20μm (Manian and Manibhushanrao, 1989).

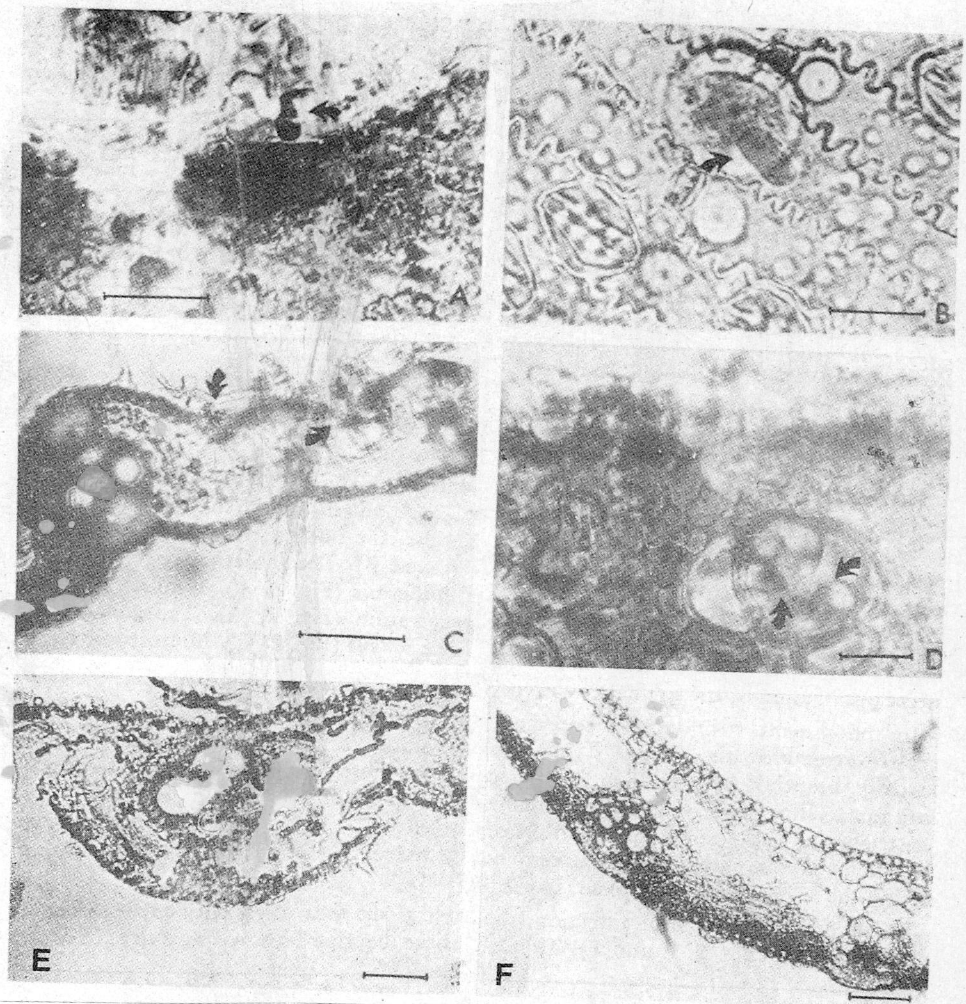

Plate 7 ABCDEF : (A) vertical section of infection cushion of *R. solani* at an early stage of pathogenesis. Arrow indicates penetration and enlargement of infection peg into the lumen of epidermal cell and browning of the surrounding host tissue, (B) epidermal leaf peeling showing the infection peg after penetration, (C) cross section through infected leaf showing the motor cells, the common sites of penetration. Arrows indicate the penetration points, (D) spread of the pathogen through vascular bundle. Arrows indicate hyphae in a vessel, (E) and (F) cross section of the leaf and sheath showing late stage of disease. Note : The disfigured and collapsed cells of rice leaves and sheath around invading hyphae. Bar = 20μm except C, where bar = 100μm (Manian and Manibhushanrao, 1989).

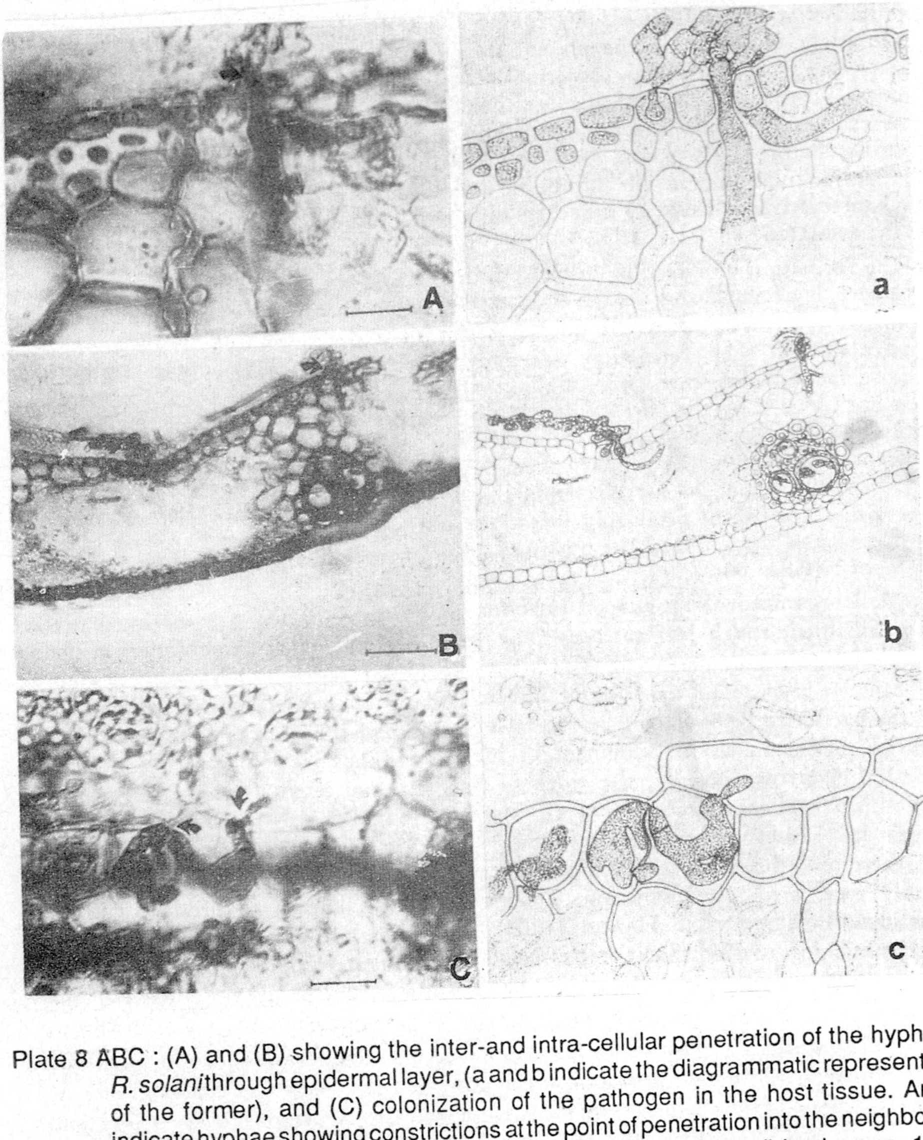

Plate 8 ABC : (A) and (B) showing the inter-and intra-cellular penetration of the hyphae of *R. solani* through epidermal layer, (a and b indicate the diagrammatic representation of the former), and (C) colonization of the pathogen in the host tissue. Arrows indicate hyphae showing constrictions at the point of penetration into the neighbouring cells and c is diagrammatic representation (Manian and Manibhushanrao, 1989).

pathogen might be the cause of extensive damage to the host tissue, where cells collapse and disfigure in the later stages of disease development.

Entry through stomatal openings

Manian and Manibhushanrao (1982b) used three rice cvs. of differential resistance to rice ShB disease such as ARC 5943 (resistant), IR 20 (moderately resistant) and IR 1487-194-5-3-2 (susceptible) to ShB disease for stomatal penetration studies by the pathogen. The pathogen entered through stomata on both surfaces of leaves and leaf sheaths. The lobate hyphae entering the stomata usually formed appressoria-like structures, almost covering the stomatal openings (Plate 6 E). On entering the stomata, hyphae occupied the substomatal cavity and from there branched and invaded the surrounding tissue. Occasionally a few hyphae re-emerged through stomata from heavily infected tissue.

The attraction of up to six lobate hyphae from different directions towards the stomata- 'stomatal tropism' was observed (Plate 6 F). The per cent stomata penetrated showed a seven fold increase in the upper leaf surfaces of susceptible (IR 1487-194-5-3-2) and a five fold increase on moderately resistant (IR 20) over the resistant cv. (ARC 5943). However, in the case of outer and inner leaf sheaths, the increase was not much in the moderately resistant and the susceptible cvs. over the resistant one (Table 6).

Table 6 : The per cent stomata penetrated by lobate hyphae of *R. solani* on rice leaves and sheaths (Manian and Manibhushanrao, 1982).

Cultivar	Leaf surface		Sheath surface	
	Upper	Lower	Outer	Inner
ARC 5943 (Resistant)	4.5	8.1	1.9	1.4
IR 20 (Moderately resistant)	24.3	30.5	3.1	3.2
IR 1487-194-5-3-2 (Susceptible)	31.0	34.5	2.5	6.0

Penetration through intact host surface

Penetration of the intact plant cuticle and epidermis of the rice leaves and sheaths is effected by the formation of infection cushions. The hyphal tip of the infection cushion produces slender infection pegs (Plate 7 A,B) from the flattened ventral surface of the cushion which was pressed against the host surface. These pegs penetrated the cuticle and cell wall (Plate 8 A,B). The infection peg regained its original size after its entry into the host epidermal cell during intracellular penetration (Plate 7 A). The common site of penetration was found to be motor cells of the leaves (Plate 7 C). In detached leaves, physical injury caused by pricking facilitated the direct penetration of hyphae without the involvement of infection cushions.

Histopathological studies of the disease revealed that the pathogen spreads on the host surface as runner hyphae before initiating infectious, lobate hyphae. Unlike the obligate parasites the pathogen with its high saprophytic ability does not need immediate entry.

Host colonization

In the sheaths, with loosely arranged parenchyma cells and air spaces, the lobate hyphae resembled the runner hyphae without much branching. However, they were repeatedly branched and swollen in the leaf tissue. Further spread of mycelia was by both intracellular and intercellular colonization. During intracellular spread, the hyphae constricted at the point of penetration of the cell wall (Plate 8 C). The leaf vessels act as barriers to spread of the pathogen to adjacent healthy tissue as seen in the cross section (Plate 7 D). In the advanced stage of infection, the whole tissue was filled with mycelia, leading to cell collapse (Plate 7 E,F).

Soon after the colonization of the tissue, the mycelium grows rapidly and initiates secondary lesions. Pectin transaminase has been implicated in pathogenesis (Lin, 1976); starch and nitrogen are known to favour disease development (Hashiba et al., 1977). The linear nature of the lesions produced by this pathogen could be due to its faster spread to healthy tissue through vascular bundles.

Ultrastructure

Earlier, Roy and Sayre (1985) used electron microscope (EM) on the mycoparasitic potential of the antagonists *Trichoderma harzianum* and *T. viride* on *R. solani*. Subsequently, Matsuura (1986) studied the infection process of ShB fungus in leaf sheaths of rice plants using scanning(S) EM. Hyphae of infection cushions grow over the epidermal cells (Plate 9A). While the hyphal tips are attached to the epidermis with a mucilage-like material around them (Plate 9B). Further, the penetration pegs developed from the infection cushion invade towards epidermis (Plate 9C,D) and their tips and pores are clearly seen (Plate 9E). Multiple invasions from epidermal cell lumen were observed (Plate 10A). Further, infection hyphae formed branched at short distance from infected point (Plate 10B). This is in confirmation with the earlier light microscopic studies made by Manian and Manibhushanrao (1982A) on *R. solani* invasion in rice leaf sheath.

Hirooka et al. (1990) reported detectable ultrastructural changes in the fungal cells for 6 h treatment of flutolanil-a systemic fungicide (2µg/mg con.) used for rice ShB control. This has resulted in sparse cytoplasm (Plate 11A,B) leading to the separation of the plasmamembrane from cell walls indicating an osmatic shock accompanied by the cell collapse (Plate 11C,D). There is a clear evidence on the destruction of cellular organelles by flutolanil. Although the reports on microscopic observations of fungal cells treated with other fungicides are available, biochemical evidence supporting cytological changes has seldom been scanty.

Flutolanil belongs to benzanilide compounds which is known to inhibit succinate dehydrogenase complex (SDC), an important enzyme complex for respiration in the mitochondrial system. Histochemically, SDC was more active *in situ* in the infection cushions than in untreated hyphae (Hirooka et al. 1990). SDC activity lead to ATP synthesis and ATP depletion may disturb the synthesis of macromolecules. In conclusion, it may be emphasized that the inhibition of succinate oxidation leads to decrease in ATP which may cause sparse cytoplasm and reduce osmatic pressure in fungal cells thereby giving rise to death of the fungus.

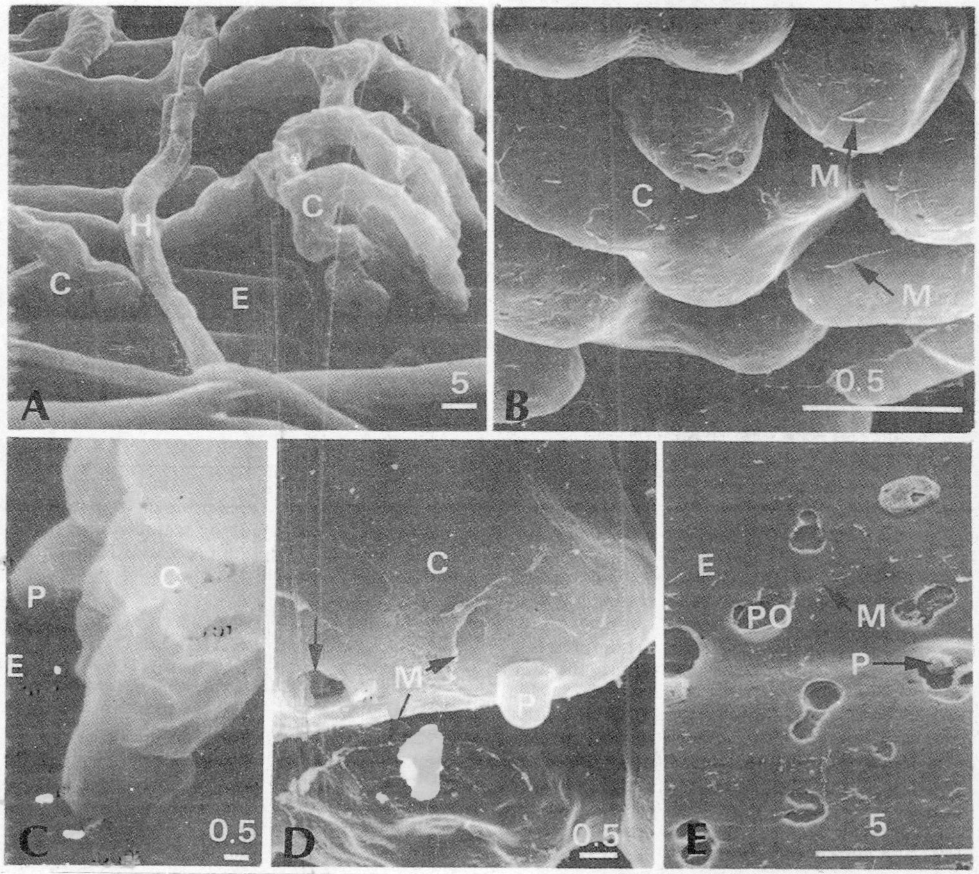

Plate 9 ABCDE : SEM studies of infection cushion formation and infection process of *R. solani* on epidermis of rice leaf sheaths. (A) hyphae of infection cushions growing over epidermis, (B) hyphal tips of infection cushion attached to the mucilagenous material around the hyphae, (C) penetration pegs extended from the basal part of infection cushion, (D) development of penetration peg from the flattened cells at the base of the infection cushion, (E) marks of the tips of penetration pegs and pores on the surface of a leaf sheath (Matsuura, 1986).

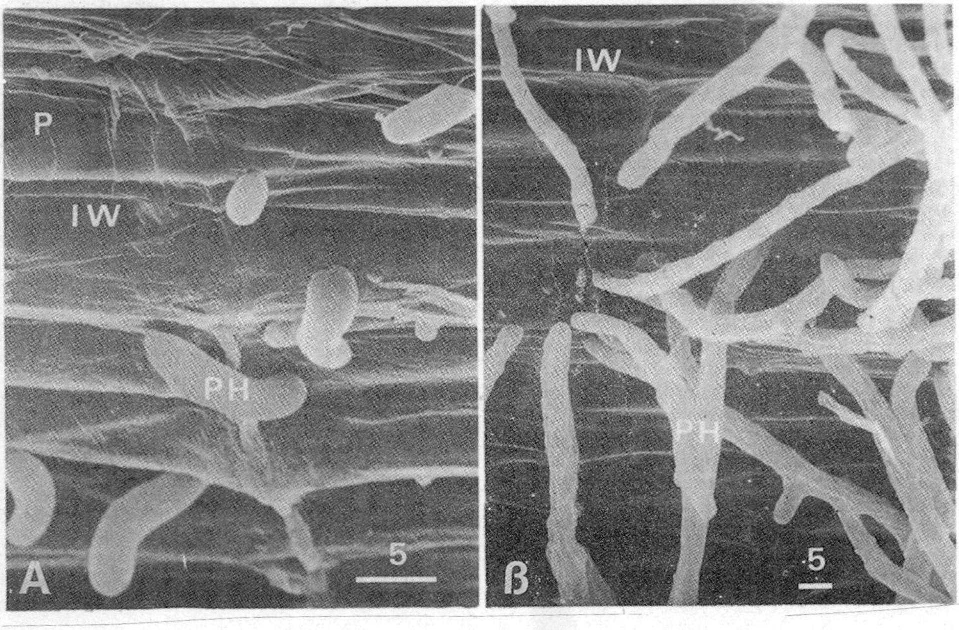

Plate 10 AB : SEM studies of penetration process of *R. solani* into epidermal cell lumen of rice sheath. (A) multiple invasion of penetration pegs into epidermal cell lumen, which was disclosed by the peeling off of the epidermal cell wall with cellophane adhesive tape, (B) infection of hyphae in a cell lumen. Infection hyphae formed branches a short distance from the introduced points. IW : infection cell wall of epidermal cell (lumen side), P : penetration peg, PH : penetration hyphae (Matsuura, 1990).

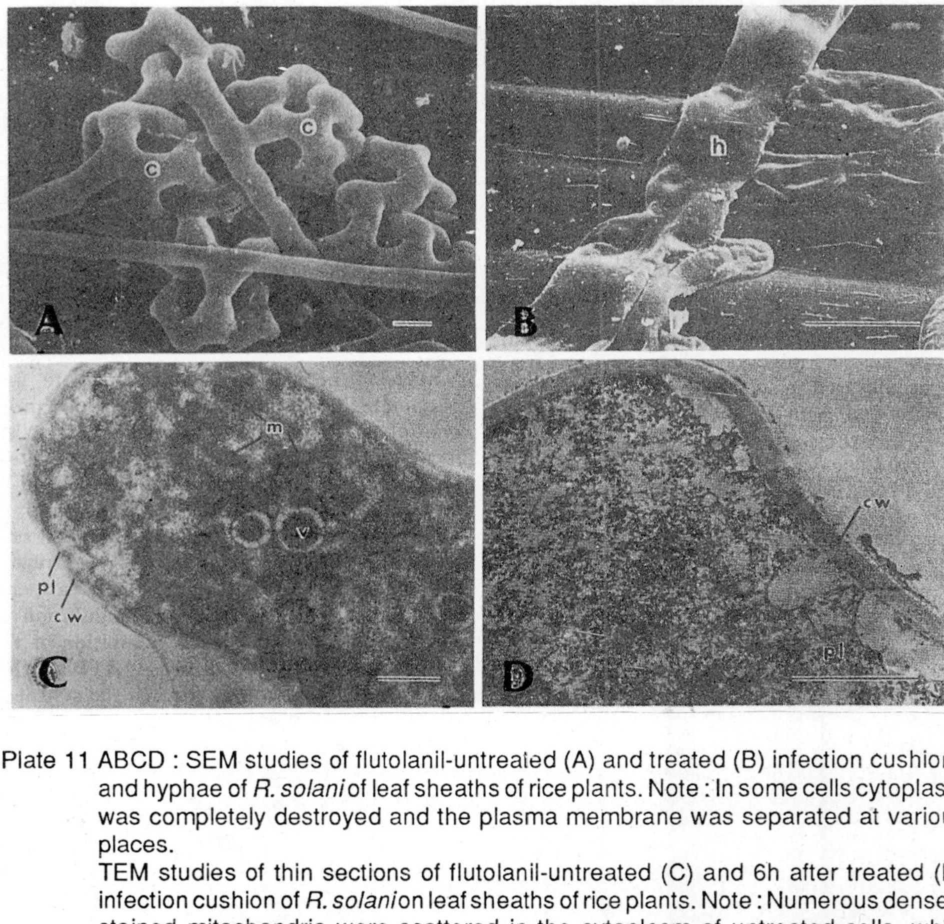

Plate 11 ABCD : SEM studies of flutolanil-untreated (A) and treated (B) infection cushions and hyphae of *R. solani* of leaf sheaths of rice plants. Note : In some cells cytoplasm was completely destroyed and the plasma membrane was separated at various places.

TEM studies of thin sections of flutolanil-untreated (C) and 6h after treated (D) infection cushion of *R. solani* on leaf sheaths of rice plants. Note : Numerous densely stained mitochondria were scattered in the cytoplasm of untreated cells, while cytoplasm was completely destroyed and the plasmamembrane was completely separated from cell wall at various places. c : infection cushion, h : flutolanil treated hyphae, pl : plasmamembrane, cw : cell wall. m : mitochondrion, v : vacuole. AB: bar = 5μm; CD: bar = 1μm (Hirooka et al., 1990).

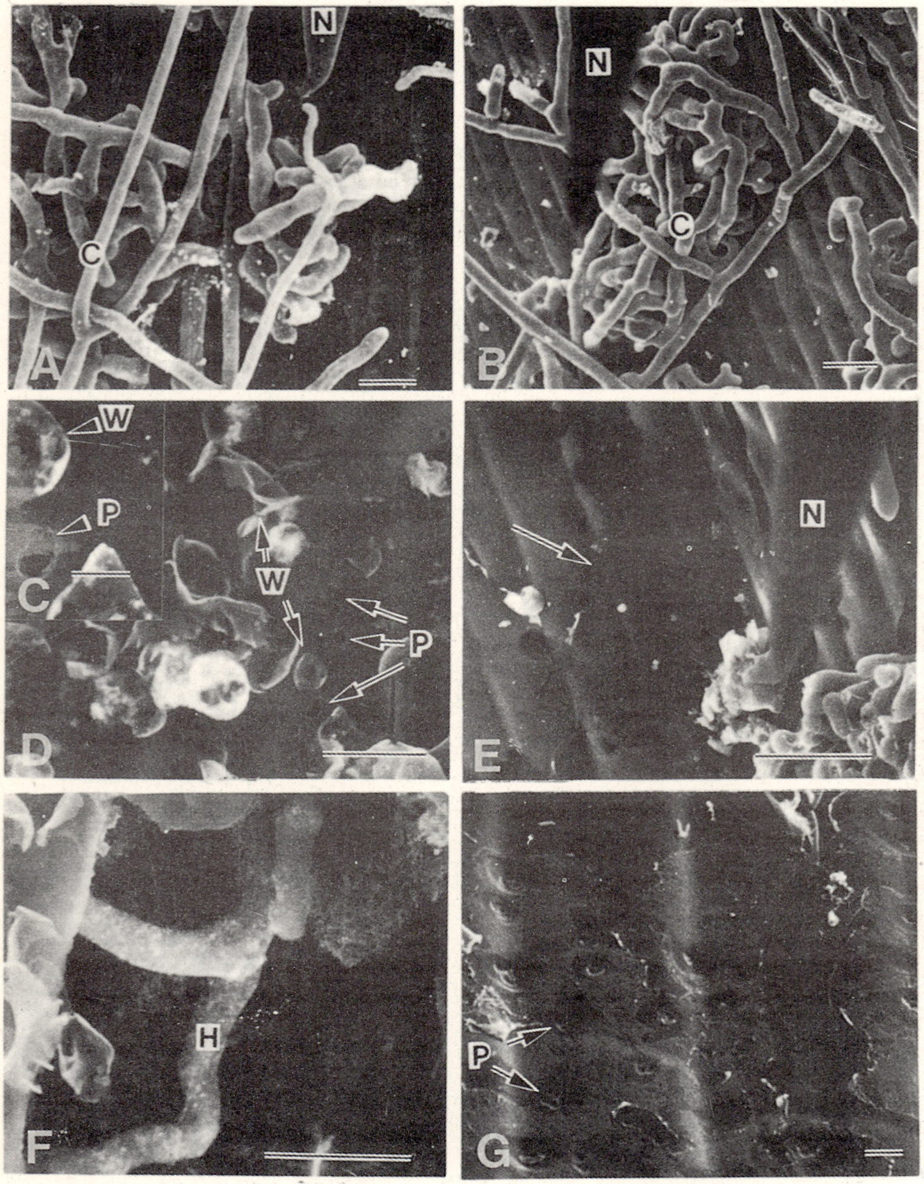

Plate 12 ABCDEFG : Micromanipulation of SEM studies of infection process of *R. solani* on the inner leaf epidermis from untreated and flutolanil-treated rice plants. (A) infection cushions on untreated leaf sheaths after 48 h after inoculation, (B) infection cushions on leaf sheaths that contained 1.95 μg/g of flutolanil 48 h after inoculation, (C) wall debris and penetration pores observed of the impression of the infection cushion. (D) enlarged portion of the impression of the infection cushion. Note : penetration pores are clearly seen, (E) infection hyphae in an epidermal cell lumen, (F) impression of the infection cushion on leaf sheath, while in (G) 48th after inoculation of flutolanil treatment (0.6μg/g). The infection cushion was removed with microneedle. Note : tiny penetration pores are visible. C : infection cushion, P : penetration pore, W : wall debris, H : hyphae, N : microneedle. ABE : bar = 10μm, CDF : bar = 5μm, G : bar = 1μm (Hirooka, et al., 1989).

The biological mode of action of flutolanil on the fungus *R. solani* and its micromanipulation through SEM studies (Hirooka et al., 1989) revealed that the per cent inhibition of hyphal growth and infection cushion formation increased as the amount of active ingredient increased. Further, SEM studies on detached leaf sheaths from untreated and flutolanil treated rice plants showed the impression of the infection cushion (Plate 12A) and the enlarged portion (Plate 12B) containing penetration pore and wall debris. The infection hypha (Plate 12C) in the epidermal cell lumen and the impression (arrow) of the infection cushion on a leaf sheath of flutolanil at 1.95µg/g con. (Plate 12D) and 0.6µg/g con. (Plate 12E) treated rice plants after 48h of inoculation was clearly seen. This study showed that the infection cushions of treated leaf sheaths at both the concentrations of the fungicide were easily removed with a microneedle as they appeared to be less firmly attached than those of treated sheaths. This study indicated that micromanipulation in an SEM is a useful technique for investigating the penetration process of *R. solani*. Further details on the effect of flutolanil on ShB are given under chemical control.

Varietal Resistance

In the absence of highly resistant cvs. against the ShB pathogen, the term *tolerance* is described as a form of general resistance or when there is no unequivocal evidence of resistance. Tolerance is considered to be the capacity to look susceptible but 'field resistant' i.e. a cv. tolerates the disease. In other words, it is a property of those cvs. which although diseased as severely as others, do not show as much reduction of yield. Here in the text the term tolerance is avoided. Although the terms HR and R were used in the screening but they could stand only for different levels of resistance to ShB disease.

Hashioka (1951 a) reported from field tests in Japan that cvs. introduced from India, Thailand, Burma, Europe and North America were more resistant than local types. He believed that resistance to ShB was a dominant character. In crosses between resistant and susceptible cvs. majority of the F_2 population was resistant and in some cases a 3:1 ratio of resistant to susceptible plants was found (Hashioka, 1951b). However, tests conducted in Taiwan (Hsieh et al. 1965), Korea (Kang et al. 1965), India (Manian and Manibhushanrao, 1982b; CRRI, 1982) showed no immune or highly resistant cvs. Reports on source of resistance to various cvs./cultures to ShB pathogen *R. solani* are available and summarized in Table 7.

Table 7 ; Summary of various rice cvs./breeding lines/cultures showing resistant reactions to ShB pathogen *R. solani* (Manibhushanrao, 1990).

Country	No. of entries	Method of screening	Disease reaction on cv./culture		Reference
			HR/R	R/MR	
Philippines	—	—	—	Kataktara DA-2	IRRI (1972)
Philippines	1368	IRSHB nursery	—	Bahagia, IR 1544-340-6, IR 3273-67P, K 8 (Mutant) Sel., Pankaj, Laka, Tapoochoz	IRRI (1976)
Philippines	—	—	Tapoochoz, Laka, Palman, BG-11-11, IR 15-29-680, IR 883-12 and IR 1542-30	—	Ou (1976)
India	36	—	—	Nijavara, Ptb 1, Ptb 8, Ptb 9, Ptb 18, Ptb 25, Ptb 27, Ptb 30, Ptb 31, Thillai-nayagam, Ratna and IET-1991	Prabhat et al. (1973)

contd ....

Country	No. of entries	Method of screening	Disease reaction on cv./culture		Reference
			HR/R	R/MR	
India	280	—	—	Chennelu, Hamsa, 1-1-4 and 11B 12	Kannaiyan and Prasad (1979a)
India	167	Stem-tape inoculation	B 1757d-Sm-6-1, B 1757d-Sm-50-6, BR 4-30-51-2, BR 51-196-2, IR 2035-224-3-2-2, IR 2070-132-2-2-5, IR 2070-747-6-6-3-2-1, IR 2071-588-4-4, IR 2071-803-5-2-6, IR 2796-44-2, IR 2798-115-2-3, IR 4422-98-3-6, IR 4531-10057-5-1-2, A 15-100-1-3-1, Suduwee, Tadukan, IET 4834 (CR 117-1), and ARC 5943	—	Manian and Manibhu-shanrao (1982b)
India	338	Natl. screening nursery	—	TNAU 17005, RNR 74822, R 245-30-41, HPU 2203, HPU 2205, HPU 2206, IET 6080 RP 1015-12-10-3-1-1, R 161-3179, KMP 8, KMP 41, RNR 82249, RP 1015-45-117-1 and RP 1015-38-85-1	Dev and Mary (1983)
India	125	Natl. screening nursery	—	OS 4, Pankaj, IR 42, IET 5656 and other 19 breeding lines	CRRI (1981)
India	—	—	—	CR 1014, T 141	Dath (1985b)
India	—	—	RP 1821-11-5-4 (OS 4 × Phalguna) RP 1122-15-2-3 (Pankaj × Swarnadhan), RP 1822-21-1-1 (Pankaj × Swarnadhan) and RP 2076-37-5-3 (Manasarovar × NLR 9674)	—	DRR (1986)
India	150	Laboratory screening		P578-157-129-18 P559-163-6-102-2 P745-1, P 1037-3-13, P615-1-1-36, P615-1-1-37, P1037-7-4, P65-1-2-39, P615-1-2-40, P1037-88-6 and P1037-88-1	Pillai (1990)

HR : Highly resistant; R : Resistant and MR : Moderately resistant

Recently, Amante et al. (1990) studied 76 wild rice accessions for the genetic resistance to ShB in IRRI's ShB nursery. Although all entries showed ShB symptoms on the leaf sheath, a few *O. minuta* (acc. 101089), and *O. rafipogon* (acc. 100907) were found to be resistant and 35 other entries were moderately resistant than the susceptible check IR 58. These accessions may be useful donors of ShB resistance in the breeding programme. However, Rutger (1990) stressed that ShB pathogen resists all skills of rice breeders.

Inheritance studies in ShB disease are meagre in the absence of resistance donors. However, Sha and Zua (1989) in China made an attempt using Tetep, IET 4699, Jawn no. 14 and Yedao as resistant donors and IR 9752-72-3-2 as susceptible parent to study the inheritance reactions using virulent isolate RH-9. The F_1 of four combinations of moderately resistant and susceptible parents showed intermediate reactions to inoculation and the F_2 distributions tend to be continuous implicating resistance to ShB is controlled by multiple genes.

Method of inoculation

Yoshimura and Nishizawa (1954) adopted various inoculation methods for testing resistance, wherein it was reported that (i) field tests using artificial inoculation increased the efficiency and uniformity of infection (ii) seedling tests did not differentiate the percentage of infection (iii) maximum tillering stage of the plant is more susceptible and optimum for varietal screening (iv) sterile straw bits inoculated with the fungus inserted among tillers of each hill and wrapping of the tillers for one week after inoculating them was most efficient and (v) sheath inoculation method, by injecting the inoculum through hypodermic needle of a syringe, similar to that used in blast disease gave no indication of any difference in resistance in the cvs. screened. IRRI carried out similar studies in nurseries with seedlings, in boxes, field inoculation in booting and flowering stages and detached leaves.

Amin (1975) suggested an improved method, 'stem-tape inoculation' for evaluating rice ShB disease by securing the inoculated stem bits firmly to the sheaths using a cellotape, 5 cm above water level. Lesions developed even at 4-d after inoculation. Guo et al. (1985) suggested uniform techniques for resistance trials. Dath (1987) proposed a multipurpose detached leaf technique using which the sclerotial viability can also be tested.

The efficacy of the scale

Ono (1953) and Yoshimura (1954) suggested scales for the assessment of ShB i.e. the degree of severity or the damage. The formula suggested by Ono is as follows:

$$\text{Degree of damage} = \frac{0A+10B+10C+25D+40E}{N}$$

where A is the number of plants without lesions, E is the number of plants in which all the sheaths are infected and B, C and D are intermediates. N is the total number of plants.

According to Yoshimura,

$$\text{Degree of severity (\%)} = \frac{3N_1 + 2N_2 + N_3 + 0N_4}{3N} \times 100$$

where N_4 is the number of tillers with 4 uppermost sheaths healthy, N_1 is the number of tillers with all the uppermost sheaths infected and N_2 and N_3 are intermediate. N is the total number of tillers.

The Standard Evaluation System (SES) for rice suggested by International Rice Testing Programme (IRTP) followed by Bandong et al. (1975) seems to be inadequate as the cvs. included in the IRSHBN trials showed different reactions when tested at different locations and seasons using the scale 0-9 units. In this scale, 0 indicates immunity, 3 indicates 21-30% of the sheaths infected, 9 indicates lesions reaching the top of the plants and all leaves showing severe infection. The flag leaf sheath infection, one of the main criteria used is hardly seen particularly in the *rabi* (dry) season. Manian and Manibhushanrao (1980) developed a scale based on a normal distribution, using lesion length, vertical and horizontal spread of the disease as major parameters from which 5 reaction types ranging (HR, R, MR, S and HS) highly resistant to highly susceptible respectively were used. The suggested scale derived from the normal distribution curve is given in Table 8. According to the formula

$$Z = \frac{(X - m)}{\sigma} \qquad X = m + z\sigma$$

where, m = mean of the reading, Z = standard normal variate, σ = standard deviation. Z values are found from the respective probabilities such as Z_1 for 0.1 (10%) probability = -1.28, Z_2 for 0.35 (35%) probability = -0.39, Z_3 for 0.65 (65%) probability = 0.39, Z_4 for 0.9 (90%) probability = 1.28. Scales were nominated as X_1, X_2, X_3 and X_4 on the basis of lesion length (cm), percentage of infected sheaths/infected tillers and percentage of infected tillers/hill and thus $\sigma - X_1$ = HR (highly resistant), $X_3 - X_4$ = S (susceptible) $X_4 - \sigma$ = HS (highly susceptible). The percentage of uppermost parts infected by *R. solani* is related to the percentage of the height of the whole plant and the degree of disease incidence. Hashiba et al. (1981) screened about 1,700 rice cvs. in Japan and observed that the vertical disease developed was more extensive in early maturing cvs. The suggested formula to estimate disease incidence was Y = 1.62 X - 32.4, where, Y = disease incidence and value of X being 0.73 Z - 4.13, Z being the height of the uppermost infected part.

Table 8 : Scale for assessing sheath blight resistance using normal distribution curve method (Manian and Manibhushanrao, 1980).

Criteria	HR ($-\alpha$- x_1)	R (x_1 - x_2)	MR (x_2 - x_3)	S (x_3 - x_4)	HS (x_4 - α)
1. Lesion length (cm)	0.71-1.17	1.18-1.72	1.73-2.20	2.21-2.75	2.76-4.00
2. Per cent infected sheaths/infected tiller	13.64-20.64	20.65-35.21	35.22-41.59	41.60.62.54	62.55-84.21
3. Per cent infected tillers/hill	3.70-9.42	9.43-20.66	20.67-30.50	30.51-41.74	41.75-77.55

Analysis of the field data especially the IRSHBN entries from different countries could be uniform even if there was a change in locality, environment and virulence of the isolate. From a field of susceptible cv. the disease seemed to spread faster to neighbouring hills than it does in the field of a resistant cv. Thus, the spread of disease both within the hills, particularly upwards, and among hills may be considered a major criteria of disease resistance (IRRI, 1972). Here, for convenience the number of lesions/tiller was omitted because this was more or less directly proportional to the number of infected sheaths/tiller. The suggested scale with five groups is more appropriate and has wider use and application as the segregation of the population is measured statistically, than the scale suggested by IRRI where the classification is mostly based on visual rating. Dath (1985b) used six parameters; lesion length, lesion type, infected area in relation to total height of the culm, severity of the leaf infection, sclerotial formation and overall build-up of the disease in the plants. Here again, a cv. which was resistant against one parameter fell into the susceptible grade when some other parameter was adopted and *vice-versa* and may lead to erroneous conclusions.

Comparison of the assessment methods

Sharma and Teng (1990b) made a comparison of five different methods viz. highest relative lesion height (HRLH%), disease severity (DS%), disease incidence (DI%) as suggested by Ahn et al. (1986), standard evaluation system (SES) for rice (IRRI, 1988) and real area infected (RAI%) for assessing ShB infection under pot culture experiments. All these methods are standard methods used in ShB assessment except RAI% where it is calculated by measuring the length and width of the infected area. Among them, per cent DS was faster than per cent HRLH and more efficient for evaluating a large amount of material within a short time. The area under the disease progress curve (AUDPC) was used to compare the efficacy of the methods on the three cvs. IR 64, IR 72 and IR 26957-86-2. All the assessment methods except DI gave the lowest AUDPC value in cv. IR 26957-86-2 confirming that it was more resistant to ShB than the other two cvs.

Physiology of Host-Pathogen Interaction

In nature 'resistance is the rule rather than the exception' (Yarwood, 1967). Every plant has some mechanism(s) for resisting most of the microorganisms in its environment. The invading pathogen has to face the host defence reactions as it enters the latter. Pathogenesis results only when the pathogen overcomes these defence reactions. Thus, the pathogen triggers host metabolism and the expression of symptom is the reflection of these alterations. Changes in the composition and amount of carbohydrates, nitrogenous and phenolic compounds of the host have often been implicated in the resistance/susceptible reactions to pathogen.

Carbohydrate metabolism

Kozaka (1961) reported that susceptibility of the leaf sheath is closely correlated with the N content, but not to the carbohydrate content especially sugar or starch. Zuber and Manibhushanrao (1983) reported the differential response to sugars in IR 20 and TKM 9 to ShB fungus infection. Total and reducing sugar content decreased in the infected resistant and susceptible cvs. This can be attributed to the fact that major part of the sugars is directed to polyphenol synthesis through shikimic acid pathway. Besides, some of the carbohydrates might have been utilized for respiration which is enhanced considerably in the diseased tissue causing a decline in the soluble sugars. Starch accumulation in the initial stages of infection and subsequent lowering can be ascribed to enzymatic hydrolysis of starch by *R. solani* (Reddy and Rao, 1978).

Nitrogen metabolism

Zuber and Manibhushanrao (1983) reported that total and free amino N decreased in both resistant and susceptible cvs. as a result of ShB infection. Initial enhancement of N in IR 20 is more pronounced than in TKM 9. However, wide fluctuations were observed in both the cvs. This corresponds with the observation made by Reddy and Rao (1978) on host N metabolism in damping-off of groundnut caused by *R. solani* infection. The total N accumulation in the later stages of ShB infection in the susceptible TKM 9 cv. is interpreted as the synthesis of more organic N compounds especially enzymatic or non-enzymatic proteins by the host plants.

Nitrogen fertilization

It has been reported that heavy application of N-fertilizer predisposes rice plants to ShB disease (Kozaka, 1961, 1965; Loo et al., 1963 and Kannaiyan and Prasad, 1979b). Munnera (1973) noticed a reduction in intensity of ShB by lowering the amount of N applied.

Manian (1982) studied the influence of N-fertilization and ShB infection on the accumulation of various cell constituents in rice plants. Split application of N to rice gives significantly better yields than single application, both through protecting the crop from severe ShB disease outbreaks and in reducing N losses through leaching and denitrification. Although both the resistant and susceptible cvs. became more susceptible with age, the differences are more prominent when they were 90 to 120-d old. There was a general increase in the concentration of free phenolics as well as the oxidative enzymes, peroxidase (PO) and polyphenol or DOPA-oxidase (PPO) in ShB infected N-fertilized plants. Reducing and total sugars, ortho-dihydroxy (OD) and total phenols significantly increased in both resistant and susceptible cvs. as a result of infection. Interestingly, these increases were more pronounced in resistant than susceptible cvs. Further, changes in enzyme activities of PAL and TAL were highly consistent with the increase of phenolics in the infected cvs. under N-fertilized and unfertilized conditions: again, the oxidising enzymes PO, PPO registered high activity (Manian, 1982).

Phosphorus and potash fertilization

Besides heavy application of N, higher quantities of P and K fertilization also tend to increase ShB incidence (Inove and Uchino, 1963). However, Rajan (1981) reported that high N and low K made the plant susceptible to ShB while least disease incidence was noticed in the lowest level of N and highest level (240 kg/ha) of K. Further, even the level of P has no significant effect on ShB incidence or grain yields.

Detection of fungal mass *in situ*

Quantitative cell wall analysis of ShB fungus revealed chitin as its major component (Potieter and Alexander, 1966). Evaluation of the mycelial mass *in situ* by assaying the cell wall chitin will be a useful method to follow the parasitic life of the pathogen. The increment of glucosamine content detected in susceptible TKM 9 and moderately resistant IR 20 ShB

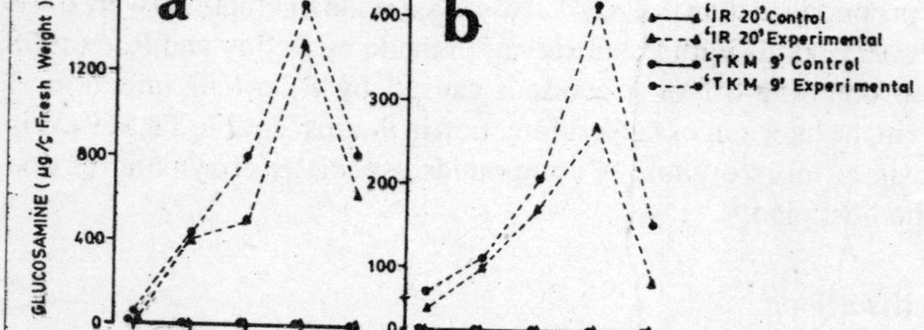

Fig. 8 ab : Glucosamine content in uninoculated and *R. solani*-inoculated rice tissues, (a) excised leaves and (b) intact plants (Manibhushanrao and Mani, 1993).

pathogen infected rice cvs. revealed greater colonization of the pathogen in the former than the latter. In order to check the suitability of the assay for glucosamine content, the quantified fungus mass was compared with two other biochemical parameters protease and free protein of the host during infection (Manibhushanrao and Mani, 1993).

All the three parameters studied showed their relationship to each other. The pace and extent of host colonization by the pathogen as evaluated by assaying the glucosamine content of the diseased tissue of excised leaves and intact plants (Fig. 8a, b) was in close proximity to the other two biochemical investigations. Glucosamine estimation as a technique to accurately monitor the fungus mass in the host tissue at any specific stage of pathogenesis of ShB fungus seems to have wide application in assessing the disease progress.

Disease Resistance

Resistant plants synthesize preformed substances, phytoalexins and other antimicrobial substances accumulated post-infectionally. The mechanism(s) of disease resistance implies broadly upon elucidating how phytopathogens damage plants and how plants limit their damage. The role of antifungal and antibacterial substances in disease resistance in cereals (Manibhushanrao and Zuber, 1978) as well as phenolic compounds and changes in the enzyme activities of phenolic biosynthesis and oxidation in host resistance against plant pathogens (Manibhushanrao et al., 1988) was reviewed.

Phenol metabolism

Kannaiyan and Prasad (1980) reported the inhibitory effect of certain phenolic compounds on the growth and sclerotial production of ShB fungus. Zuber and Manibhushanrao (1984) subsequently monitored the changes in phenolic contents of resistant (IR 20) and susceptible (TKM 9) cvs. of rice at different stages of ShB disease development. Contrary to the expected rise in phenolic content in the resistant plants, OD phenols accumulated earlier and faster in TKM 9 than in IR 20. Total phenols increased similarly in TKM 9. Resistant and susceptible cvs. failed to present a consistent pattern in the accumulation of total phenols and flavonoids. Again, flavonoids increased in IR 20 only in the later stages of infection. It is difficult to explain the marked fluctuations noted in the accumulation of these cell constituents in the diseased tissue. However, reduction in sugar content associated with an increase in phenolics in the TKM 9 suggested that major part of the sugar is shunted to polyphenol synthesis.

Phenol biosynthetic enzymes

Zuber and Manibhushanrao (1984) reported that in rice seedlings inoculated with ShB fungus, there was decreased activity of key phenol biosynthetic enzymes PAL and TAL except in the susceptible cv. TKM 9 where there was an increase in TAL activity (Fig. 9 a, b). No PPO activity was detected in uninoculated or inoculated plants. PO decreased in both the rice cvs. as a result of infection and the decreased enzyme activity are highly variable with enhanced phenolic levels in ShB fungus—inoculated rice plants.

Changes in phenol constituents

Ohata et al. (1966) observed that phenolic acids especially *p*-coumaric and ferulic acids with flavonoids confer resistance in rice plants by arresting the growth of the invading pathogen. Kannaiyan and Prasad (1980) reported the inhibitory effectof *p*-coumaric and ferulic acids on the growth and sclerotial production of ShB fungus *in vitro*. Manibhushanrao et al. (1986) studied the changes in phenolic constituents of ShB fungus in infected rice tissue. A general decrease in the concentration of soluble phenols and an increase in that

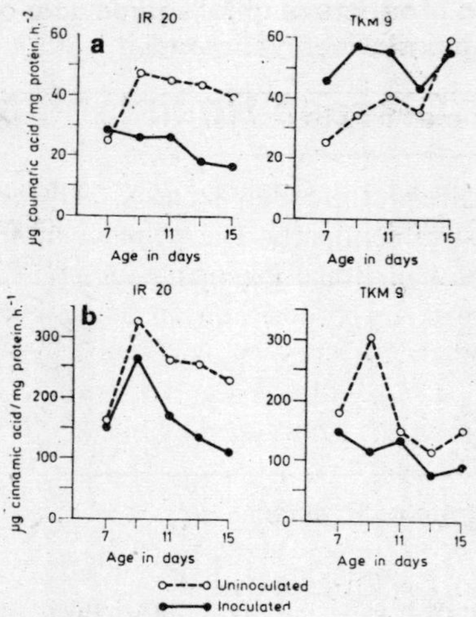

Fig. 9 : PAL (a) and TAL (b) activity in uninoculated and *R. solani*-inoculated resistant (IR 20) and susceptible (TKM 9) rice seedlings (Zuber and Manibhushanrao, 1984).

of bound phenols were observed in the diseased tissue of both susceptible and resistant IR 20 cvs. TLC fractionation of the bound phenols revealed the presence of two monohydric phenols namely, *p*-coumaric and ferulic acid and four other unidentified compounds. Quantification of *p*-coumaric and ferulic acids revealed their enhanced level, and can be attributed to a possible role in the host resistance against *R. solani* infection.

Effect of PAA on phenol metabolism

Manibhushanrao et al. (1989) studied the effect of PAA on the phenolics and enzymes of phenol metabolism in rice cvs. IR 50 and TKM 9, moderately resistant and susceptible to ShB respectively. PAA at low concentrations (0.1 and 1.0 ppm) resulted in the rapid accumulation of phenolic compounds in both the cvs. Accrual of total and OD phenols was greater in TKM 9 and the flavonoids was higher in IR 50. Further, PAA elicited the activity of PAL, TAL, PPO and PO in both IR 50 and TKM 9 cvs. Interestingly, the magnitude of increase of total and OD phenols and the activity of PAL, TAL and PO was also greater in TKM 9 than that of IR 50.

Oxidized products of chlorogenic acid

Polyphenols are known to inactivate pectic enzymes by their oxidized products such as quinones. Earlier Manibhushanrao and Sreenivasaprasad (1985) studied the effect of chlorogenic acid and its oxidized products on polygalacturonase (PG) production by *R. solani*. The quinone of chlorogenic acid inhibited the PG activity by about 47% and this could be attributed to the oxidation of SH - groups of the enzyme by the quinone or the formation of 1,4 additional compounds by the quinone with the amino group of the enzyme. Table 9 shows the percentage inhibition of PG activity by the oxidised products of chlorogenic acid at different concentrations of the phenol.

Table 9 : Effect of oxidized products of chlorogenic acid on PG activity of *R. solani* (Manibhushanrao and Sreenivasaprasad, 1985)

Concentration in μm/ml	Enzyme activity (units/ml)		Inhibited activity (units/ml)	Percentage inhibition
	Phenol + phenolase	Original*		
0.5	13.42		3.24	19.4
1.0	11.58		5.08	30.5
		16.7		
1.5	8.85		7.81	46.9
2.0	10.25		6.41	38.5

* PG activity in the culture filtrate of *R. solani.*

Penetration and invasion of host tissue by a pathogen is promoted by the production of pectic enzyme(s) and/or toxin(s). Inactivation of pectic enzymes by oxidised products of polyphenols is well known and may be a mechanism responsible for inhibiting the pathogen attack. However, such an assumption requires confirmation from studies involving either pectinase or toxin or by the synergistic action of both of them.

Induced resistance

Inducible type of resistance mediates through augmenting the synthesis of constitutive inhibitory compounds (prohibitins) and/or by the formation of phytoalexins following infection and their accumulation to toxic levels. The phytoalexins are antimicrobial substances formed as a result of host-parasite interaction and may play a crucial role in the biochemical defence of plants against infection. Literature on germinated seeds of several plants challenged by the native microflora producing phytoalexins is available. Zuber and Manibhushanrao (1979) reported production of a phytoalexin-like substance by the germinated paddy seeds challenged with ShB fungus. The partially purified, antifungal TLC fraction markedly inhibited the radial growth of ShB fungus as well as other test organisms. Recently, acquisition of resistance to ShB fungus in response to biotic (Kalaiselvi et al., 1986; Manibhushanrao et al., 1990) and abiotic stress such as PAA and its derivatives (Waheeta et al., 1987; Manibhushanrao et al., 1989; Sarkar and Sinha, 1989) has been reported.

Phytotoxin pretreatment

Manibhushanrao and Ramalingam (1993) found that PAA and its derivatives induced typical lesions on susceptible TKM 9 rice leaves at high concentrations (100 ppm and above). Waheeta et al. (1987) reported that at very low concentrations, (0.1 and 1.0 ppm) PAA and its derivatives induced partial resistance in pretreated excised rice leaves (about 25-30% protection was achieved) (Table 10). None of the metabolites tested at these concentrations were either phytotoxic or fungitoxic. The fact that only specific low concentrations of PAA and its derivatives are able to impart resistance in rice and that concentrations only slightly higher induced susceptibility further rules out the possibility of any direct action of phytotoxins on the pathogen. Presumably, the acquisition of resistance in rice to ShB is due to the physical and/or chemical barriers elicited in the host by the phytotoxin pretreatment.

Table 10 : Effect of PAA, *m*-HPAA, *o*-HPAA and *p*-HPAA treatment of rice leaves on sheath blight spread (Waheeta et al., 1987).

Treatment	Concn. (ppm)	Disease severity* (%)	Per cent protection
Control		100	
PAA	0.1	103	-3
	1.0	70	+30
	5.0	155	-55
	10.0	200	-100
m-HPAA	0.01	98	+3
	0.1	70	+30
	1.0	120	-20
o-HPAA	0.01	105	-5
	0.1	73	+28
	1.0	78	+23
p-HPAA	0.01	100	-
	0.1	75	+25
	1.0	125	-25

*Based on disease severity index scored 5-d after R_5 inoculation

Host sensitization

Chemical sensitization of host tissue offers promise as an alternative to classification strategies used for protection of rice from ShB. Synthesis of phytoalexin in plant tissues can be stimulated by elicitors that are produced by plant pathogens and by several factors that are apparently unrelated to pathogenesis such as UV light, freezing and chemical substances such as, polyamines, antibiotics, DNA intercalating agents and salts of heavy metals (Anderson-Prouty and Albersheim, 1975).

Manibhushanrao et al. (1990) studied a variety of chemicals such as lithium sulphate, sodium azide, sodium selenite (heavy metal salts); thioglycollic acid (reducing agent); cycloheximide (antibiotic-metabolic inhibitor) and penicillin (antibiotic-growth regulator) as elicitors in susceptible rice cv. TKM 9 against the highly virulent isolate R5, (ATCC No 48505). The phytotoxicity and fungitoxicity of the test chemicals were assayed (at 10^{-3} to 10^{-6} M) to determine per cent seed germination and mycelial growth of ShB fungus *in vitro*. Maximum protection (49%) was achieved with thioglycollic acid (10^{-4} M) followed by sodium selenite (44%) at 10^{-4} M and lithium sulphate (38%) at 10^{-3} M (Table 11). Leaf diffusate from rice seedlings when bioassyed against the radial growth of ShB fungus and conidial germination of *Drechslera oryzae*, (incitant of brownspot disease of rice) thioglycollic acid treated infected rice tissue gave 71 to 83% inhibition respectively. The ShB spread was reduced by thioglycollic acid and sodium selenite at 10^{-5} M yielding 51 to 40% protection respectively. Sarkar and Sinha (1989) also identified a number of phytoalexin induced chemicals to control rice ShB. The elicitation of resistance has been attributed to the formation of antifungal compounds in response to infection. These elicitors might be more effective to protect for a shorter duration against seedling (ShB infection) blight which occur in susceptible rice cvs. in the nursery as well as in upland conditions.

Table 11 : Comparative effect of chemical inducers on rice (TKM 9) against *R. solani* infection (Manibhushanrao et al., 1989).

Treatment		% infected seedlings	
		Seed treatment	Seed+root-dip treatment
Water (control)		90 gh	90 f
Lithium sulphate			
	10^{-3} M	56 bc (38)	56 cd (38)
	10^{-4} M	76 ef (16)	64 e (29)
Sodium azide			
	10^{-4} M	73 de (19)	46 b (49)
	10^{-5} M	78 ef (13)	58 cde (36)
Sodium selenite			
	10^{-4} M	50 ab (44)	—
	10^{-5} M	58 c (36)	36 a (60)
Thioglycollic acid			
	10^{-4} M	46 a (49)	—
	10^{-5} M	62 c (31)	35 a (61)
Cycloheximide			
	10^{-5} M	64 cd (29)	—
	10^{-6} M	75 ef (17)	62 de (31)
Penicillin			
	100 ppm	79 efg (12)	49 bc (46)
	500 ppm	82 efg (9)	—
	1,000 ppm	85 gh (6)	—

—plants dried; Values in columns followed by different letters differ significantly (P=0.05); Values in parenthesis denote per cent protection

Incompatible interaction

Waheeta et al. (1987) reported that incompatible host parasite interaction (prior inoculation) of rice leaves reduced the ShB spread considerably. Concomitant inoculation of the leaves with virulent R5 and avirulent R7 (from potato) resulted in a 52.4% reduction in the disease incidence whereas the magnitude of resistance was clearly greater in rice leaves prior inoculated (48 h) with R7 yielding 81% protection (Table 12). The challenger (R5) could not literally spread till 3 d and only from 4 d a few lesions appeared. Further, Manibhushanrao et al. (1988) reported that after 12 h of treatment by incompatible interaction (12 h) of excised rice leaves with rice-grain culture of the avirulent isolate (R7),

Plate 13 ABCD : Cross protection in rice leaves against ShB, (A) cv. Ponni leaves inoculated with *R. solani* isolate R5 (control), (B) cv. Ponni leaves prior interaction with R7 and R5, (C) cv. IR 50 leaves inoculated with R5 (control), (D) cv. IR 50 leaves prior interaction with R5 and inoculated with R5 (Manibhushanrao, et al., 1988).

significant protection was achieved (90%) against differentially virulent ShB fungus isolates (IR and R5). The challenger (pathogenic) inoculum could not spread till the third day. Both the susceptible Ponni and resistant IR 50 displayed high degree of acquired resistance to R5, resulting in marked reduction of ShB incidence (Plate 13 ABCD). In the absence of direct antagonism and physical obstruction, nutritional competition and/or induced host resistance appear to be the plausible mechanisms involved in the rice - ShB fungus system.

Table 12 : Effect of avirulent isolates interaction of rice leaves on sheath blight spread (Waheeta et al., 1987).

Treatment	Disease severity (%)	Per cent protection
Control (R_5)	100	-
$R_5 + R_7$	47.6	+52.4
R_7 (48 h) + R_5	19	+81

*Based on disease index scored 6-d after R_5 inoculation.

Dath (1988) studied the infectivity and variability of different inocula of ShB fungus isolates in upland rice and found a general reduction in severity of the disease. This might be attributed to the incompatible interaction among the isolates used as inocula.

The mechanism(s) of action to confirm resistance by the antimicrobial substances is much less studied and proper answers to whether phytoalexins play a major role in determining plant resistance are some of the questions which require careful investigation. Similarly, whether *in vitro* or greenhouse studies on host sensitization and induced resistance by abiotic and biotic agents, reproducible under field conditions giving maximum protection to the rice plants are also worth considering.

Genetic basis of host resistance

Genetic analysis of many fungal plant pathogens has established that genes for race specificity, often called virulence/avirulence genes are encoded in the nucleus. Comparatively, a few genetic studies on many other genes that contribute to pathogenicity also support the above view. There has been little evidence to suggest that cytoplasmic genes could have been overlooked if they were present in the cytoplasm of all the strains used in genetic crosses.

Recent evidence shows that cytoplasmic determinant is required for virulence in *R. solani* (Flinker et al., 1985). All virulent isolates examined so far contained several dsRNA (double strand RNA) segments, atleast some of which were encapsidated in isometric particles, whereas hypovirulent isolates had either no dsRNA or fewer dsRNA segments. The virulent factor could be transferred along with the dsRNA segments, from a virulent strain (resistant to benomyl, sensitive to BTN fungicide, melanin producing and with large sclerotia) to a hypovirulent strain (sensitive to benomyl, resistant to BTN, no melanin produced, small hyaline sclerotia) by hyphal anastomosis, giving a virulent strain possessing all the nuclear markers of the hypovirulent strain. Hence virulence factor was reasoned to be cytoplasmic. Proof of the role of one or more specific segments of dsRNA in virulence will require infection and/or transformation of protoplasts with isolated dsRNA. In this respect it is noteworthy that Martini et al. (1978) have suggested a DNA plasmid which may be required for the pathogenicity of *R. solani* and Hashiba et al. (1984) reported the presence of a plasmid DNA designated as $_{p}$RS64 in *R. solani* which was weakly pathogenic. These results do suggest that there is a possible relationship between extrachromosal DNA and pathogenicity.

Stability of horizontal resistance

Van der Plank (1968) attributed stability of horizontal resistance to stabilizing selection because of interaction of a large number of genes. The ShB pathogen population responds to horizontal resistance in the cvs. by accumulating genes for virulence, not genes for pathogenicity. This proposition was evaluated at IRRI, in ShB disease, and the incitant for which, in rice, no monogenes for resistance are known.

Crill et al. (1981) tested more than 20,000 selections from the IRRI rice germplasm bank to evaluate their reaction to ShB disease. Among the least affected were cvs. Tapoochoz, Bahagia and Laka. Cultivar IR 1487-372-1-1 was considered susceptible and compared with the mean per cent infection using 8 isolates (CO - C7) of ShB fungus isolated from these 4 cvs. for 7 successive crops (Table 13). Simultaneously, isolates obtained from a single cv. for 7 crop cycles were tested against all the 4 cvs. (Table 14). Disease severity caused by isolates obtained from the susceptible cv. either did not change or slightly decreased on the above three horizontally resistant as well as on the susceptible one. These results indicate that ShB pathogen could respond tc the selection pressure applied by horizontal resistance causing increased disease severity and this increase might be presumed to result from gradual accumulation of polygenes which, collectively, govern virulence. The pathogen seems to be unstable but is dynamic, and a change depends on the selection pressure applied.

Table 13 : Comparison of the mean per cent infected leaf sheaths 3 weeks after inoculation with *Rhizoctonia solani* isolates obtained from 7 crop cycles (C1-C7) on 4 cvs. with the original (Co) isolates. (Crill et al., 1981).

R. solani isolates from different crop cycle	Infected leaf sheaths (%)			
	Tapoochoz (horizontally resistant)	Bahagia (horizontally resistant)	Laka (horizontally resistant)	IR 1487-372-1-1 (susceptible)
Co	34.7	39.6	30.8	55.1
C1	50.0	33.1	38.6	35.6
C2	30.4	37.1	31.2	46.2
C3	31.2	32.9	29.1	32.3
C4	32.7	40.7	36.3	65.4
C5	52.0	33.8	16.7	49.3
C6	27.9	74.3	31.2	42.8
C7	62.6	60.7	36.5	40.7

Table 14 : Comparison of the mean percentage of infected leaf sheaths of 3 horizontally resistant rice cvs. and susceptible cv. inoculated with *Rhizoctonia solani* isolates obtained by successive isolation and reinoculation of 7 consecutive crop cycles of the same 4 rice cvs. (Crill et al., 1981).

		Infected leaf sheaths (%)			
		Horizontally resistant to *R. solani*			Susceptible
Cultivar	Crop cycle	Tapoochoz	Bahagia	Laka	IR1487
Tapoochoz	C1	28.6	32.9	25.9	39.3
	C2	32.1	35.6	37.0	53.9
	C3	37.2	34.6	35.4	46.7
	C4	24.1	25.0	23.1	33.7
	C5	30.9	30.5	32.5	47.4
	C6	19.7	33.7	43.2	70.5
	C7	38.0	37.2	37.7	58.9
Bahagia	C1	28.6	23.8	23.4	35.4
	C2	28.2	34.2	35.5	54.2
	C3	40.2	40.0	38.7	68.8
	C4	25.0	32.9	33.3	51.5
	C5	31.7	30.5	29.1	36.4
	C6	42.5	86.9	54.5	71.2
	C7	44.9	54.7	59.4	81.4
Laka	C1	34.7	31.7	40.3	45.7
	C2	36.4	40.5	35.5	71.4
	C3	42.3	46.7	41.0	62.8
	C4	20.5	32.0	22.8	26.0
	C5	32.9	34.1	35.4	38.3
	C6	26.0	32.0	28.8	54.8
	C7	27.3	34.2	35.1	36.8
IR 1487	C1	29.3	30.9	36.8	33.8
	C2	22.5	33.8	25.0	28.4
	C3	22.6	21.7	26.0	27.3
	C4	29.6	29.6	27.5	45.2
	C5	32.9	37.7	21.5	20.2
	C6	32.0	30.0	32.0	54.2
	C7	30.4	26.9	24.6	29.7

Disease Control

Kozaka (1975) studied ShB and its control in rice plant. In the tropics where yields have been low, the general recommendation can be aimed by planting high yielding resistant cvs. as well as on cultural practices such as spacing and N application (Roy, 1978). ShB incidence can also be prevented by using spray schedules with potent chemicals as well as antibiotics but it is an expensive proposition.

There have been many studies on the ecology and control of ShB under temperate conditions in Japan (Hashiba, 1985a) and in Korea (Kim et al., 1981, 1985, 1988; Yu, 1977, 1981). There appears to be some psuedo-sheath blight pathogens which are associated with the ShB creating problem for its control. Lee and Courntney (1982) recommended burning rice straw which compliments ShB control tactics. In the tropics where yields have been low, cultural practices such as spacing and low N application are envisaged (Roy, 1978).

Chemical control

The chemical control on rice ShB in Japan was studied by Yoshımura (1984) and intensively in various prefectures namely from Chiba (Igushi, 1988) and Oita (Hasama, 1989). In recent years, the total area treated with chemicals for the control of ShB in Japan has reached 1.7 million ha (Hashiba and Ijiri, 1989). Korea alone spends every year 2-3 million kg of fungicides to control ShB.

A major breakthrough in the chemical control of diseases is the development of systemic fungicides. However, the use of several of these toxic chemicals has been problematic due to their undesirable side effects such as phytotoxicity, carcinogenicity, teratogenicity, residual and pollution effects.

Organoarsenicals such as methylarsine sulphite, methylarsine-bis-dimethyl dithiocarbamate (Hashioka, 1956; Hashioka and Makino, 1960; Chu and Chen, 1964), ferric methyl arsenate (Takita et al., 1965) were most effective in killing the sclerotia. Combined sprays (containing ziram and methylarsine-bis-dimethyl dithiocarbamate) at 0.1-2% (Abeygunawardena and Desilva, 1966) and halogenophenylacetyl hydrazine derivatives (Sumina et al., 1966) reduced the disease incidence effectively. Organic tin compound (10%) triphenyl-tin acetate, as wettable powder with 85% kaolin and surfactants, controlled the disease in pot/field tests at concentration as low as 1:1000 by arresting the ShB fungus growth (Tamura, 1965).

Chien and Chu (1973) reported that benomyl absorbed by the host tissue was very effective against ShB pathogen infection while the soil drenching fungicide hymexazol (3-hydroxy-5-methylisoxazole) was a good eradicant and protectant against ShB (Takahi et al., 1974). Hymexazol in aqueous solutions (50 ppm) induced a marked decrease in the mycelial dry weight, and leakage of mycelium protein which showed a linear increase with the increase in concentration of the fungicide until 250 ppm and then levelled off (Nakanishi et al., 1975).

The fungicidal control of ShB disease in India was attempted by Kannaiyan and Prasad (1976) and Bhaktavatsalam et al. (1977). The rhizosphere population of ShB fungus of rice seedlings was drastically reduced through foliar sprays of the fungicides such as kitazin, edifenphos, benomyl, carboxin and carbendazim. The efficacy of benomyl (Das and Panda, 1984) carbendazim (Bhaktavatsalam et al., 1977; Rajan and Alexander, 1988) in the management of ShB was studied. Benomyl and captan at 0.2% were highly effective in reducing the seedling infection while soil drenching with edifenphos, kitazin and benomyl during tillering stage was also effective in controlling ShB. Seed treatment with carbendazim, chloroneb, chlorothalonil, carboxin, benomyl and phenyl mercury acetate (PMA) reduced the seed-borne infection of ShB fungus, improved the seed germination, shoot and root growth, seedling vigour and prolonged the viability of the seeds. Again, 0.2% sprays of benomyl, kitazin, edifenphos and chlorothalonil were highly effective in controlling ShB and increased the grain yield in field trials. Roy and Saikia (1976) obtained the best control of ShB with carbendazim or by benomyl sprays (0.05% concentration) both in greenhouse and field tests. In field trials with six fungicides, kitazin granule was the most effective in reducing the disease severity, followed by edifenphos (Varma and Menon, 1977) but Mathai and Nair (1977) showed that edifenphos was the best as it increased the yield.

Bandong et al. (1978) tested several fungicides as sprays on rice ShB control. The results from the two trials showed that both formulations used at the rate of 3 or 1 kg/ha application of iprodione produced the best control as well as the highest grain yield. Others were fairly good tc ShB control include mebenil, chlorothalonil, edifenphos, TCP, kitazin and zinc omadine.

The occurrence and chemical control of ShB in Korea have been reviewed (Yu, 1981). Moncern and neoasazin were tested on the development of ShB on susceptible cv. Jinheung under greenhouse conditions and found the former too had long duration, more then 30-d from the time of application (Kim and Kim, 1988). Further, Kim et al. (1989) studied the effect of Neoasazin Lq on the development of sclerotial diseases of rice which includes ShB. Suryadi and Kadir (1989) tested 8 fungicides against ShB control in the field and triazole was most effective in inhibiting the spread of ShB lesions. Significant lesion reduction also occurred with chlorothalonil, mancozeb and maneb + zineb treatments.

Flutolanil, a new systemic fungicide developed with both protective and curative properties is very effective to control various *Rhizoctonia* group of fungi including rice ShB (Araki, 1985; Mochizuki et al. 1987; Hirooka et al., 1988, 1989). However, ShB disease was effectively controlled by 80% at low concentration of 1.6 to 3.2 μg/g plant. Foliar spray, soil drenching (submerged application) and seed treatment have been tried effectively under greenhouse or field studies. Yamada (1986) reported a non-systemic fungicide pencycuron with specific effect against rice ShB fungus as well as other plant disease caused by *Rhizoctonia* spp. Both the wettable powder and dust formulations are remarkably stable and was not adversely affected by the rainfall after the chemical treatment to prevent ShB disease spread (Kuck et al., 1988). Diclomezine, yet another potent fungicide with its long lasting efficacy, controls several sclerotial diseases of rice including ShB (Takahi et al.,

1988). Foliar application at panicle initiation, to heading stage at the rate of 160-480 g a.i./ha gave excellent ShB control. Bicyclic benzanilides (Oshumi et al., 1989) and aromatic ketone pyrimidinyl hydrazones (Konishi and Kuragono, 1989) were reported to be effective against ShB.

In Japan, forecasting the incidence of ShB was based on the number of viable sclerotia which was closely correlated to disease severity and timely application of fungicides 15-d before heading and booting should be more effective (Yamaguchi et al., 1971 and Hori and Anraku 1971). Earlier Chien et al. (1969) made several such studies in Taiwan.

The sclerotia and propagules of *R. solani* are tolerant to various adverse conditions in soil. Reports on the inhibition of sclerotium formation with sulfur -containing amino acids such as cystine, cysteine, homocysteine, taurine and methionine are available (Moromizato et al., 1980 a,b; Tazaki et al., 1982). Other sulfur containing compounds; phenylthiourea, phenylthioethanal, phenylthioacetic acid, dithiothreitol and thioglycollic acid were also inhibitive (Zuber et al., 1984). Further, the inhibitory effect of 40 derivatives of 1, 3-dithiol synthesized compounds was tested on the growth and sclerotial formation of various ShB fungus isolates (Matsuyama et al., 1985). These results are based mostly on *in vitro* studies and further critical work is needed for the evaluation of these compounds for the effective control of ShB.

Weedicide PCP (pentachlorophenol), has also been found to control ShB (Ono and Iwata, 1961; Takatsu and Nishimura, 1962; Inoue and Uchino, 1963; Endo et al., 1965). Application of the PCP and other PCP-NA herbicide in the irrigation water prevented sclerotial germination and initiation of infection structures (Hashioka and Makino, 1961; Endo et al., 1965).

Besides, reports on certain other herbicides/insecticides such as propanil, thiobencarb, chlorpyriphos and phorate (25-500 µg/ml) were found to inhibit the sclerotia/mycelial growth of ShB fungus, under *in vitro* and field conditions (Dath and Swain, 1979; Vasavan et al., 1980; Varma et al., 1980; Lakshmanan and Nair, 1980). In most studies, emphasis was given to the individual effects, but not to the combined effects of the pesticides. However, Singh and Sethunathan (1987) studied the individual and combined effect of the pesticides of different origin either simultaneously or in rotation, to reduce the mycelial growth and germination of sclerotia of ShB fungus. Of the several pesticides studied, the insecticide fenitrothion and herbicide thiobencarb were most effective in inhibiting mycelial growth and sclerotial formation at 7.5 to 20 µg/ml concentration, while the fungicides carbendazim and tolclofos-methyl showed almost complete inhibition at 2.5 µg/ml (Table 15). Interestingly, a synergistic increase in toxicity to mycelial growth, sclerotial formation and germination was noticed when the fungicide mancozeb and the herbicide thiobencarb were applied in combination at subtoxic concentration.

Although agro-chemical applications are considered as pollution hazards, it is essential to develop a strategy in which the potent chemicals are seen in the whole context of needs of the farmer. Major concern should be on quality control, testing the probability of development of resistance, safety guidance to overcome or break resistance and a thorough knowledge on the safe and economic use of the pesticides. Further, efforts on the improvement of new scientific formulations emphasizing more on multi-protectant chemicals which should play a vital role in both biotic and abiotic stresses by regulating the plant growth should be developed for effective control and enhanced yield potential.

Table 15 : *In vitro* effect of different pesticides on mycelial growth and sclerotial formation in *Rhizoctonia solani* (Singh and Sethunathan, 1987)

	Inhibition (% over control) of mycelial growth at pesticide concentration (µg/ml)								
Pesticides	1	2.5	5	7.5	10	20	30	50	100
Fungicides									
Carbendazim	56+	100	100	100	100	—	—	—	100
Tolclofos-methyl	87+	91	96	96	96	—	—	—	100
Mancozeb	0+	24+	72+	93+	93+	—	—	—	100
Insecticides									
Fenitrothion	0+	53+	76	89	92	—	—	—	100
Endosulfan	0+	20+	31+	66+	67+	74+	78	84	90
Chlorpyriphos	0+	38+	51+	58+	60+	76+	77	80	93
Phorate	—	—	—	—	0+	0+	10+	25-	87
Dichlorovos	—	—	—	—	—	—	—	—	27+
Carbofuran	—	—	—	—	—	—	—	—	0+
Herbicides									
Thiobencarb	0+	0+	0+	22+	60+	82	95	100	100
Butachlor	0+	22+	33+	40+	69+	79+	87	89	92
Control	0+	0+	0+	0+	0+	0+	0+	0+	0+

+ Sclerotia formed
- Not determined

Antibiotics

Antibiotics neohumidin (Yamamoto et al., 1965) and the D fraction of the antifungal polyoxin complex (Sasaki et al., 1969) were effective against ShB fungus. Thirumalachar et al. (1969) studied *in vitro* activity of three antifungal antibiotics against three phytopathogenic fungi including ShB fungus. Kannaiyan and Prasad (1978b) reported that aureofungin reduced the disease severity of ShB. Validamycin (Isawa et al., 1971 and Furuta, 1973; Yamamoto, 1985) and Polyoxin developed in Japan have been found effective against *Rhizoctonia* group fungi. The former was extensively tested against ShB fungus by other workers (Wakae and Matsuura, 1975 and Eguchi et al., 1973; Kim and Kim, 1988) and later showed fairly good inhibitions to ShB fungus (Bandong et al., 1978) and its efficacy was very promising. Validacin-A (3% liquid/0.3% dust), earned extremely favourable reputation all over the world and registered as effective controlling agent for ShB (Japan, 1972; Korea, 1976; Taiwan, 1977; Columbia, 1977). It inhibited mycelial growth and sclerotial formation by direct contact (Shigemoto et al. 1989) further, the surrounding of the treated lesion turned brown and did not spread any more. Chingfengmeisu and Jinggangmycin (respectively, Shanghai Institute of Plant Physiology, 1975 and Shanghai Institute of Agricultural Pesticides, 1975), from China were effective for rice ShB control as well as other crop diseases. Many others like bleomycin, baridamycin (3% solution and 0.3% dust) showed with varied degrees of success to control ShB in field trials.

Biological control

Biological control has attained importance in modern agriculture to curtail the hazards of extensive use of toxic chemicals for pest and disease control. This approach is self sustaining, efficient, and may not be long term in action. However, it plays pivotal role in modern disease management programmes. Despite impressive research on biocontrol in recent years, there are no fungal microbial agents for plant disease control registered in the U.S. except two each in Europe and the U.K.

Some important relevant terms may be considered here. A clear distinction between pathogen and parasite are quite essential. A *parasite* lives in or on another organism from which it obtains nutrient while a *pathogen* is a parasite/saprophyte which produces disease in the host. Again, *antagonist* is defined as the microorganism used in the biocontrol of plant disease. The antagonist adversely affects another organism growing in association with it. In fact anatagonist is considered as the balance wheel of nature. Further, *antagonism* is considered as the ability of the antagonist to suppress the pathogen and operates chiefly through, competition, parasitism and antibiosis. Much work has been done on understanding the mechanisms of these three major components by which the biological agents reduce the impact of pathogen populations.

Competition is the phenomenon where the biocontrol agents and the plant pathogens compete for space, nutrients, even for a particular substance(s). For instance, fluorescent pseudomonads produce some extracelluar 'siderophores' (microbial iron chelating agents) which complex the iron and make it less available/unavailable to the pathogen. During competition a successful biocontrol agent will maintain its population at higher levels to provide protection to roots against pathogens.

Parasitism is an ability of an antagonist to parasitize on other pathogens. On the other hand *mycoparasitism* (=*hyperparasitism*) is the parasitism of one fungus by another. Several of the soil-borne hyperparasites have been exploited in biocontrol studies in recent years. Hyperparasitism includes different kinds of interactions like coiling of hyphae around the pathogen, hyphal fusions, penetration, production of haustoria like structures as well as lysis of hyphae. Hyperparasites are *nectrotrophic* where they kill the susceptible host cells by means of toxins, antibiotics and/or cell wall degradings enzymes viz. chitinases, β-1, 3-D gluconases and proteases.

Antibiosis is the mechanism by which an antagonist releases an antibiotic or other volatile substance(s) which inhibits growth of the pathogen in the soil. In most cases it is demonstrated in *in vitro* studies than in *in vivo* condition.

In recent years integration of compatible, biological, chemical and physical methods has yielded promising results. Biocontrol of rice ShB through the use of fungal antagonists, organic amendments as well as antagonistic plant associated bacteria (PAB) gained momentum in view of their ability to control the disease under greenhouse as well as field conditions. Although it may be less effective but more viable when compared with chemical pesticides (Harman, 1988).

Antagonistic bacteria

In the beginning antagonistic bacteria were recognized as the key factor in biological control of ShB disease as they were found to be potential antagonistic agents. Namai et al. (1985)

identified a bacterium which produce substances having antifungal activity against phytopathogenic fungi including the ShB fungus. Seed treatment or spray application with *Pseudomonas aurofaciens* reduced the disease incidence and increased the grain yield at IRRI (IRRI, 1988). Earlier, IRRI (1985) initiated greenhouse studies on bacterization of rice seeds (cv. IR 36) with a non-fluorescent bacterium (In-b-17) and reported a considerable reduction in the ShB incidence. Mew and Rosales (1984, 1986, 1992) further reported that rice ShB was suppressed by seed bacterization and spray application of bacterial antagonists to control ShB which led to an increase in grain yield. Such beneficial effect of fluorescent or non-fluorescent bacteria isolated from South Indian rice fields were also antagonistic towards *R. solani* (Vasanthadevi et al., 1989). In the greenhouse tests, pretreatment of sclerotia in bacterial suspensions resulted in reduction in ShB lesion size up to 31 to 44% in IR 20 and 58 to 74% in TKM 9 rice (Table 16). In field plots, IR 50 and TKM 9 rice plants raised from bacterized seeds had 65 to 72% less ShB than those plants from untreated seeds (Table 17).

Table 16 : Effect of antagonistic bacteria on sheath blight lesion size in greenhouse experiment (Vasanthadevi et al., 1989).

Rice cultivar	Treatment of sclerotia	Sheath blight lesion length (mm)	Reduction in sheath blight (%)
IR 50	NF2 - treated, 1 d	16.5	19.2
	NF2 - treated, 1 wk	1.4	93.1**
	Not treated (check)	20.4	
TKM 9	NF2 - treated, 1 d	84.6	39.5**
	NF2 - treated, 1 wk	29.9	78.6**
	Not treated (check)	139.9	
ADT 36	NF2 - treated, 1 d	74.2	28.2*
	NF2 - treated, 1 wk	36.3	60.3**
	Not treated (check)	103.3	

Each value is a mean of 100 measurements:
Calculated "t" value significant at P = 0.05(*), P = 0.01(**) and not significant(ns).

Table 17 : Effect of seed bacterization on sheath blight lesion size in field experiment (Vasanthadevi et al., 1989).

Cultivar	Seed treatment	Sheath blight lesion length (mm)	Number of tillers per hill	Grain yield per plot (kg)
IR 50	Pfcp - treated	30.0**	17	2.5*
	NF2 - treated	31.1**	17	2.5*
	Not treated (check)	109.0	11	2.2
TKM 9	Pfcp - treated	32.4**	20	2.7*
	NF2 - treated	31.0**	15	2.4
	Not treated	94.0	10	2.2

Each value is a mean of 100 measurements:
Calculated "t" value significant at P = 0.01(**), P = 0.05(*) and not significant(ns).

Further, the fluorescent bacteria were readily detected after seed bacterization on rice plant parts such as root, coleoptile, first and second leaves even after 45-d sowing (Mew, 1992). Detailed quantitative studies on the population dynamics of the introduced antagonistic plant associated bacteria (PAB) against ShB disease is available (Dorairajan, 1993). Among the three PAB strains introduced to the transplanted seedlings in the field, strain 6854-8 gave significant reduction of the ShB disease. Sheaths at mid-tillering, maximum tillering, panicle initiation and early booting stages showed PAB colonization and it was more with the sheaths than on leaves (Fig. 10ab). These studies provide information, especially on the dosage and spray schedule as well as weather parameters that influence the PAB colonization of the rice system in relation to ShB disease.

Recently, Vidyasekaran et al. (1995) developed a formulation in which the bacterium *P. fluorescens* survived for a long time without any loss in efficacy. This formulation as pot based culture at 10g/kg seed as seed treatment effectively controlled rice ShB and almost two fold increase in the yield (Table 18) in several field trials. Further, Vidyasekaran and Muthamilan (1995) have claimed that this product can survive more than 8 months and compatable with fungicides need for seed treatment.

Bacterial strains which show strong antagonism as well as production of cellulolytic enzymes *in vitro* studies are potential orgnaisms for control of plant diseases. However, Thara and Gnanamanickam (1994) reported lack of correlation between chitinase production by bacterial antagonists and suppression of ShB disease.

Table 18 : The efficacy of *Pseudomonas fluorescens* in the management of rice sheath blight under field conditions (Vidyasekaran et al., 1995).

Treatment	Rice sheath blight disease severity (ShB HRLH)	Yield Kg/ha
Psedomonas fluorescens	7.0 (16)	7155
Control	20 (27)	4000

The value in parenthesis are angles; angles = $\sqrt{(\text{percentage})}$

Treatment	CD(P=0.05)
ShB HRLH	2.2
Yield	1133

Antagonistic fungi

Among the many potentially antagonistic soil fungi, *Gliocladium* and *Trichoderma* spp. of late, have been used as biocontrol agents as they have been termed as presumptive mycoparasites. Mycoparasitism of rice ShB fungus, *R. solani* by *Trichoderma* spp. (Ogura and Akai, 1965; Hashioka and Fukita, 1969; Roy 1977a; Rosales and Mew, 1982 and Roy and Sayre, 1984); *Aspergillus terreus* (Roy, 1991); *Gliocladium* spp. (Gokulapalan and Nair, 1986) and *G. virens* and *T. longibrachiatum* (Manibhushanrao et al., 1989ab) is well known.

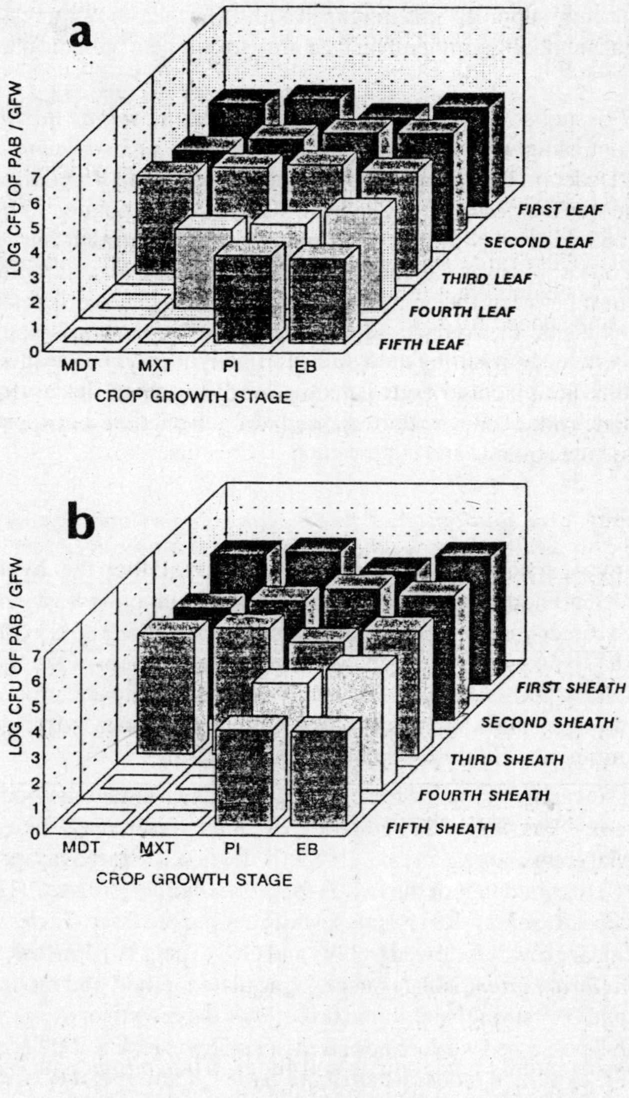

Fig. 10ab : Population of PAB strain 6854-8 on different (a) leaf and (b) sheath positions at different growth stages of transplanted rice. MTD: mid-tillering; MXT: maximum tillering; PI: panicle initiation and EB: early booting stages (Dorairajan, 1993).

Aluko and Hering (1970) studied the mechanism associated with the antagonistic relationship between *Corticium solani* and *G. virens*. The presence of *Neurospora crassa* in the soil suppressed the *R. solani* propagules. However, the former did not affect the growth of rice seedlings suggesting the possible use as biocontrol agent (Endo et al., 1973). IRRI (1987) reported that *Trichoderma* spp. isolates from rice fields were common in upland than in low land soils. With high competition for plant residues, they might exhaust the nutrient supply and thus may suppress ShB fungus in soil. However, the suppressive effect might be through fungal metabolites rather than its ability to compete with the pathogen for nutrients.

Hadar et al. (1979) studied the suppression of pathogen *R. solani* in soil causing damping off of egg plant using wheat bran culture of *Trichoderma harzianum*. Recently, Van der Boogert and Deacon (1994) reported the success of *Verticillium biguttatum*, a potential mycoparasite on *R. solani*, the incitant of black-scurf of potato. *V. biguttatum* which cannot utilize cellulose or N-nitrogen, required biotin for growth but grows and sporulate on *R. solani* on cellulose as sole organic carbon, utilizing N and biotin from the pathogen. In dual culture between the antagonist and test pathogen, the former produce haustoria-like branches inside the hyphal cells of the host. The haustoria supported an external mycelial network without killing the host cells. Later, the mycoparasite sporulates and the infected host cells die as it is ascribed to the creation of a nutrient sink by the parasite. Suppression of sclerotial production was also reported when the mycoparasite infected the localized region of the host colony.

Mycoparasitic potential

Studies on the mycoparasitism of the antagonists reveal that the hyphae of *T. longibrachiatum* form small hook/peg like structures pressing on to the host, coil loosely around the host hyphae directly or produce small branches that coiled tightly (Plate 14A). On the other hand, with *G. virens*, the host hyphae intermingled, certain of the host hyphae cells which were vacuolated became empty. Further, the hyperparasite coil around the host hyphae and adhere through small appressoria-like structures (Plate 14B) which even penetrate them at certain points (Manibhushanrao et al., 1989b).

Quantitative assessment of mycoparasitic potential of the two antagonists on PDA and SM-I (synthetic medium, Okon et al., 1973) did not reveal any preference for a particular nutritional source (Manibhushanrao et al., 1989b). Besides, Sreenivasaprasad and Manibhushanrao (1990) reported that of the two *T. longibrachiatum* isolates, Tl1 revealed better colonization against *R. solani* (R9) isolates, while among the three *G. virens* isolates Gv2 exhibited higher linear growth followed by Gv1 and Gv3 (Table 19). Further, antibiosis of *G. virens* and *T. longibrachiatum* isolates using agar plate method and block bioassay showed that both Gv1 and Gv3 strongly inhibited (80-100%) the growth of *R. solani* isolates R9 and R10 whereas Gv2 possessed weak or no activity (13-38%). Isolate Tl1 failed to show any antibiosis, while Tl2 exhibited moderate attribution (48-57%) against R9. In the block bioassay Gv1 and Gv3 revealed clear inhibition zone of 6-8mm against R9 and R10 isolates. Similarly, isolate Tl2 showed zones of 3-9mm against R9 (Table 20). Toxic metabolites production accounted as per cent inhibition on isolate R9 by *G. virens* and *T. longibrachiatum* was optimum. Under shake culture, potato dextrose broth (PDB) on 6-d (71.4 and 42.8% respectively) and Gv1 possessed markedly greater activity than Tl2 (Fig. 11 abc).

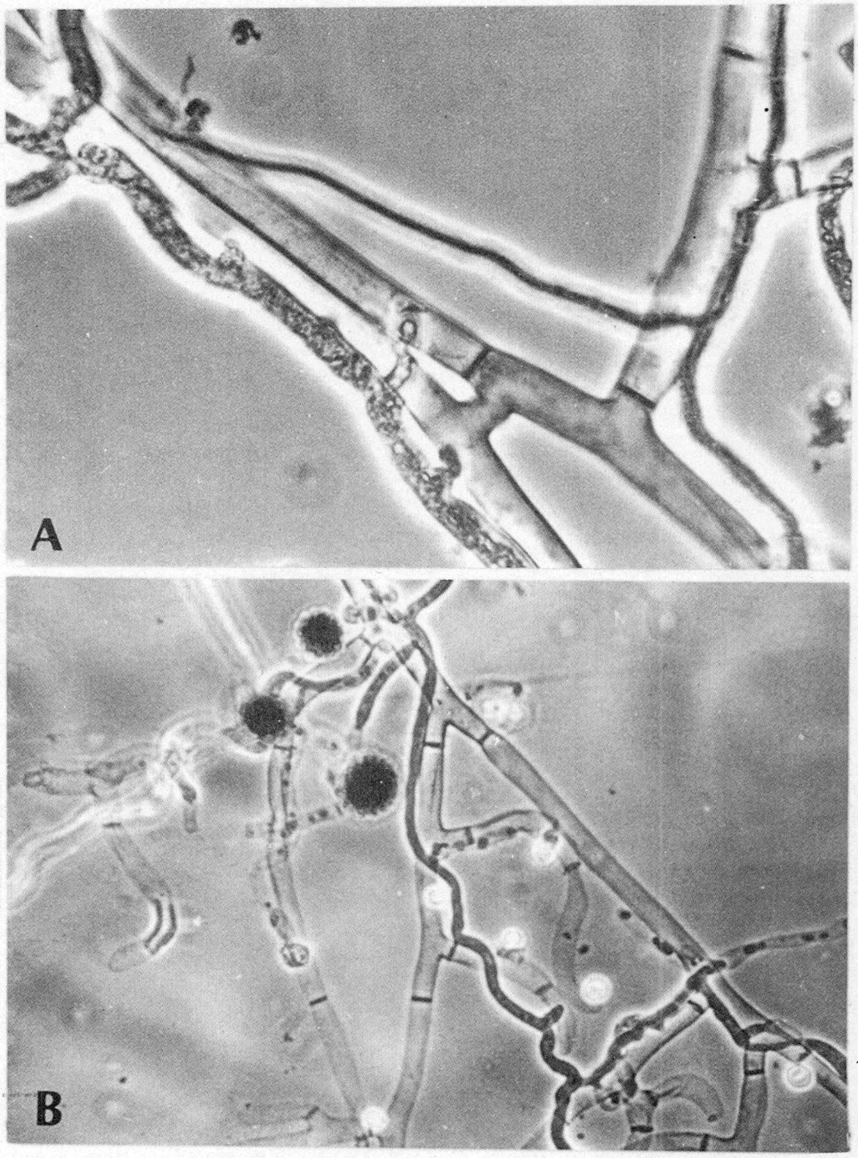

Plate 14 AB : (A) morphological interaction between *R. solani* and the antagonist *Trichoderma longibrachiatum* hyphae attached to the host by producing hooks/ pegs (X 500), (B) *Gliocladium virens* hyphae running along the host forming small appressoria like intrusions (X 340) (Manibhushanrao, et al. 1989).

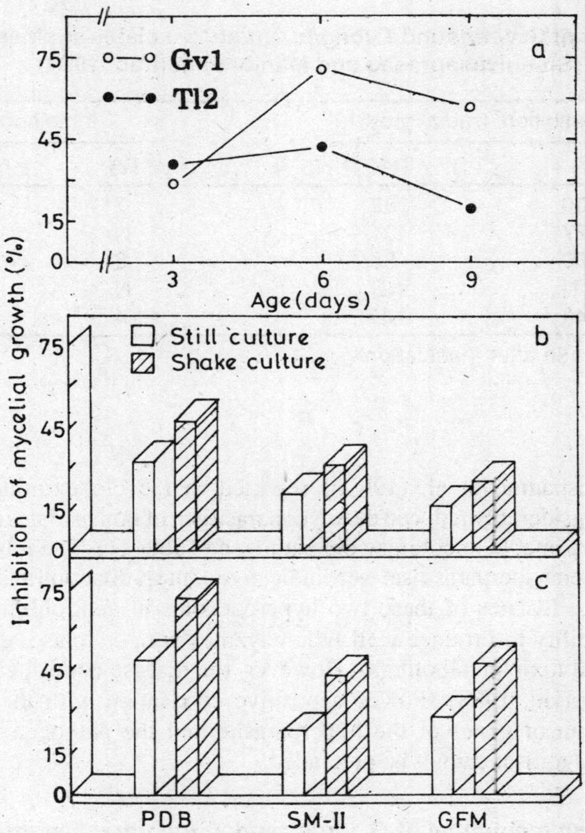

Fig. 11abc : Toxic metabolite(s) produced by *G. virens* and *T. longibrachiatum*; (a) effect of age, (b) and (c) in three different media, still and shake culture by Tl2 and Gv1 respectively. PDB: Potato dextrose broth; SM-I : Synthetic medium; GFM : Gliocladium fermentation medium. (Sreenivasaprasad and Manibhushanrao, 1990)

Table 19 : Hyperparasitic potential of *G. virens* and *T. longibrachiatum* isolates against *R. solani* (R9) (Sreenivasaprasad and Manibhushanrao, 1990).

Antagonist	Linear growth (mm)			% Plate area*		
	Time (h)					
	48	96	144	48	96	144
Gv1	27[c]	62[a]	71[b]	65[a]	82[a]	99[ab]
Gv2	28[c]	62[a]	76[a]	65[a]	81[ab]	100[a]
Gv3	20[d]	48[b]	70[bc]	36[b]	53[d]	88[c]
Tl1	36[b]	47[bc]	66[c]	62[a]	75[bc]	97[b]
Tl2	42[a]	44[c]	59[d]	64[a]	72[c]	90[c]

Values followed by different letters differ significantly.

Table 20 : Antibiosis of *G. virens* and *T. longibrachiatum* isolates against *R. solani* (R9 and R10) (Sreenivasaprasad and Manibhushanrao, 1990).

Antagonist	% inhibition* (radial growth)		inhibition zone (mm)@	
	R9	R10	R9	R10
Gv1	100	82	7	8
Gv2	27	31	N	N
Gv3	100	88	8	6
Tl1	N	ND	N	ND
Tl2	48	ND	4	ND

*@ Recorded 72 and 48h after inoculation
N—Nil
ND —Not determined

Earlier, Manibhushanrao et al. (1989b) reported that cycloheximide-the protein synthesis inhibitor, considerably reduced the mycoparasitism of both isolates at 10 ppm and at 25 ppm as they were unable to colonize the pathogen (Table 21). The inhibitory effect of cycloheximide on their hyperparasitism were indicative of the production and involvement of wall lytic enzymes. Isolates of these two hyperparasites in dual culture showed no correlation in their ability to produce wall lytic enyzmes viz., cellulase, glucanase and chitinase (Fig. 12) and toxic metabolite(s). However, the disease control efficacy of the isolates (Sreenivasaprasad, 1988) showed a positive correlation with these attributes. Again, the involvement of genes of the hyperparasite and the pathogen in regulating different levels of antagonism cannot be precluded.

Table 21 : Mycoparasitic potential of *G. virens* and *T. longibrachiatum* isolates and the effect of cycloheximide (CH) (Manibhushanrao et al., 1989b).

Antagonist	Linear growth (mm) in dual culture, Time (h)						Linear growth† (mm)		
	48		96		144		Control (SM-I)	SM-I with 10 ppm CH	SM-I with 25 ppm CH
	P*	S**	P	S	P	S			
Tl1	36.5	28.5	47.0	50.5	60	63	71	54.0	40.0
Tl2	37.0	30.0	50.0	54.0	60	61	69	41.5	25.5
Gv1	28.0	24.0	62.5	60.0	73	72	78	58.0	36.5
Gv2	28.0	23.0	60.0	62.0	71	70	76	53.5	32.0
Gv3	17.0	20.0	48.0	51.0	65	69	67	46.0	29.0

* PDA; ** SM-I; † Recorded on day 8.

Baby (1992) further studied the biocontrol potential of *G. virens* (Gv1) by incorporating into soil the wheat-bran-saw dust (WBSD) preparation (15-d) at 10 to 25 g/kg soil and obtained 10 and 57% protection respectively against ShB incidence on susceptible TKM 9. Incidentally, isolate Gv1 showed strong inhibition to the pathogen

Plate 15 ABCGv : Isolate of *G. virens* (Gv) in WBSD preparation at 25g/kg gave significant increase in growth of rice seedlings (B) than control (A). Strong inhibition to *R. solani* by antagonist (Gv) isolate to the pathogen. Control plate (C) shows no inhibition (Baby 1992).

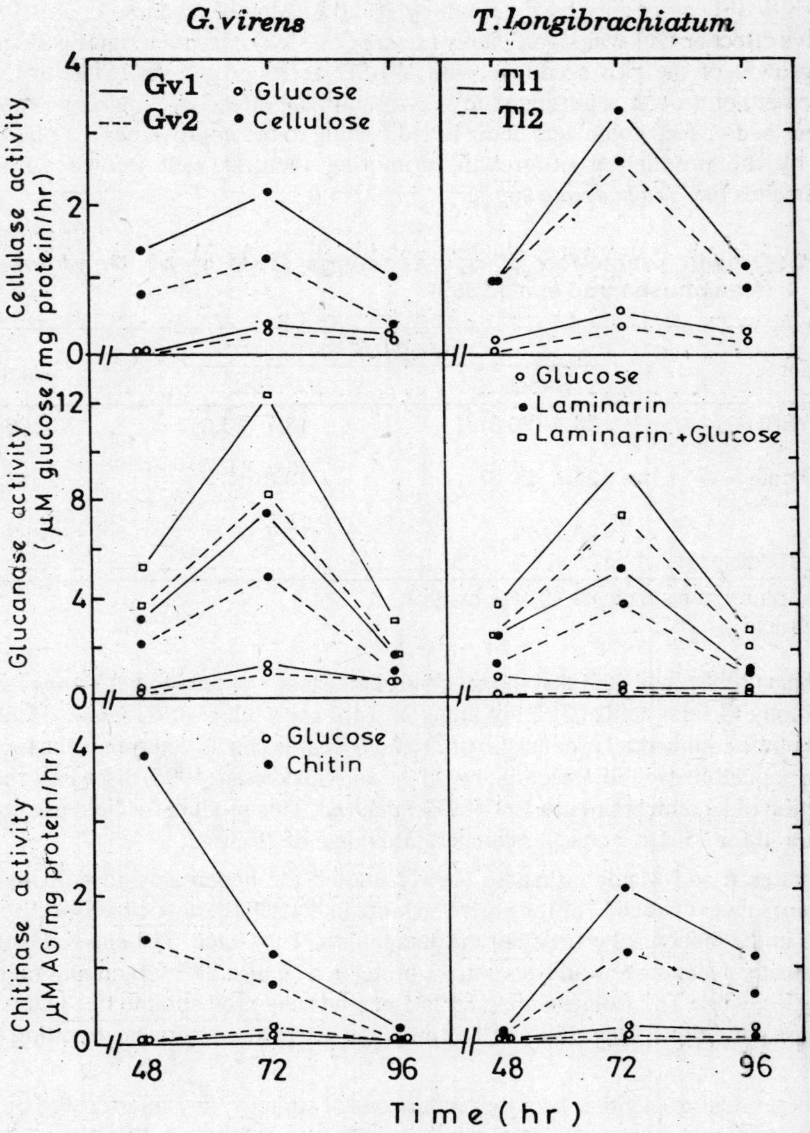

Fig. 12 : Production of wall lytic enzymes by *G. virens* and *T. longibrachiatum* isolates (Sreenivasaprasad and Manibhushanrao, 1990).

R. solani (Plate 15 AB). Quantification of biomass at 25 g/kg soil level increased the shoot and root length as well as biomass (98% increase of net dry weight) (Table 22). Although WBSD alone increased growth by 10-20% (Manibhushanrao et al., 1989) the promotive effect of Gv1 was significantly greater (73-83%), there is a significant increase in the growth of the rice seedlings with WBSD at 25 g/kg level (Plate 15C, Gv.). Enhancement of growth is attributed to 'phytosanitizing effect' of *G. virens*—the minor pathogens and/or seed colonizers being killed leading to the improvement of plant health and/or by the production of growth promoting factors (well known with other hyperparasites like *Trichoderma* spp.).

Table 22 : Growth promotion of rice seedlings (TKM 9) by *G. virens* (Gv1) (Manibhushanrao et al., 1989)

Treatment	Length (cm)		weight (mg)
	Shoot	Root	
Gv1 + WBSD (25 g)	34.4 (70.3)	13.7 (28.0)	3366 (83.6)
WBSD alone (25 g)	24.4 (20.8)	10.8 (N)	2033 (10.9)
Control (healthy seedlings)	20.2	10.7	1833

Values in parenthesis indicate % over control
N = Negligible

Adams (1990) indicated that the biocontrol efficiency of antagonist *Trichoderma* as $5x10^4$ colony forming units (CFU) of antagonist for every CFU of *R. solani*. Hence, the quantity of the commercial products(s) of *Trichoderma*/ha can be determined based on the *R. solani* populations in soil. Recently, Jeyarajan and Nakkeeran (1995) discussed about the exploitation of a commercial product of *Trichoderma*. This product for field use could be safely stored for 75-d at a constant temperature range of 20-30°C.

Kumaresan and Manibhushanrao (1991) studied the hyperparasitic potential of *T. harzianum* isolates (Th1 and Th2) against *R. solani* to understand the associated mechanism(s) involved in the biocontrol efficacy of the antagonists. The isolate Th1 showed extensive hyperparasitism leading to wall lysis and the protection against ShB fungus was more in *in vitro* studies, while Th2 isolate surpassed Th1 in rendering protection to rice plants under greenhouse conditions. This ability of Th2 might be due to more aggressive coiling of host hyphae.

It is pertinent to mention here that in biocontrol strategy, any interference by direct introduction of antagonistic microorganisms with the ecological balance of the soil microflora might not yield the expected result over a long term experiment. However, the ecological balance of soil saprophytes can be manipulated by a suitable organic amendment to promote a specific group of soil microflora.

Protoplast isolation and fusion

Earlier, mycelial protoplasts from *R. solani* were isolated and purified by Hashiba (1982b) Hashiba and Yamada (1982) using an enzyme mixture which consisted of cellulase, macerozyme and β-glucuronidase. The intact protoplasts were further separated from mycelial fragments and cell debris using aqueous two phase system made of each 0.6 M sucrose and mannitol.

Mrinalini and Lalithakumari (1995) reported strain improvement using protoplast fusion technology for *Trichoderma harzianum* and *T. longibrachiatum* a potent antagonist and tolerant to copper sulphate and carbendazim respectively. The fusants exhibited enhanced antagonistic potenital against several phytopathogens including rice ShB pathogen *R. solani*. They have also reported the increased growth response as well as ShB control using fusants (hybrid strains) under pot culture experiments.

Soil amendments

The role of organic amendments in the suppression of *Rhizoctonia* spp. has been emphasized by several workers (Papavizas and Davey, 1960; Loc et al., 1963; Rajan and Menon, 1975; Rajan, 1990; Kannaiyan and Prasad, 1981b; Dath, 1982; Kundu and Nandi, 1985; Padmakumari and Balakrishnan, 1987 and Manibhushanrao et al., 1989). Further, Central Rice Research Institute, India, (CRRI, 1980) and Dath (1982) reported the effect of green manuring on the viability of sclerotia in soil and found that the viability of the sclerotia was drastically reduced in the soil amended with green manures viz. dhaincha (*Sesbania aculeata*), sunhemp (*Crotalaria juncea*), mung (*Cicer arietinum*). However, *Azolla* sp. and *Ipomea cornea* did not affect the viability of sclerotia. Roy (1985b) reported the inhibitory effect of dhaincha on rice ShB pathogen.

Further, Baby (1992) studied the influence of amendments such as gliricidia leaf, neem cake, and tannery waste on the microflora of rice field soil infected with ShB fungus in relation to disease control in greenhouse trials. Gliricidia leaf and neem cake suppressed the pathogen growth significantly as the per cent colonization was 10 and 14 respectively (Fig. 13).

Manibhushanrao et al. (1989a) reported the efficacy of nine organic substances namely oil cakes, cotton (*Gossypium herbaceum*), neem (*Azadirachta indica*), groundnut (*Arachis hypogea*), gingelly (*Sesamum indicum*) and illupei (*Bassia longifolia*); agricultural/ industrial wastes, rice (*Oryza sativa*) husk, coconut (*Cocos nucifera*) pith and green manures, gliricidia (*Gliricidia maculata*) and crotalaria (*Crotalaria striata*) leaves as soil amendments in reducing the survivability of ShB fungus propagules. Survival of the pathogen and the population dynamics of the native microflora were monitored during 0 to 60-d decomposition period of amendments. The study indicated that dhaincha stimulated the soil microflora remarkably followed by coconut pith and neem cake while others to a lesser extent (Fig. 14abc).

The upsurge in the total microbial population was most conspicuous by 15-d followed by slight decrease on d-30 and 40, and a marked increase by d-60. Among the three groups of microbes monitored, bacterial populations (Fig. 14c) were triggered to maximum by dhaincha. Interestingly, dhaincha, coconut pith and neem cake have increased actinomycete populations (Fig. 14b) right from d-15 decomposition while most amendments did not enchance fungal populations (Fig. 14a) except cotton and neem on 30-d onwards. The efficacy of the ShB pathogen can be attributed to the direct inhibitory effect as well as their ability to trigger the residential microflora particularly the actinomycetes. Thus, reduction

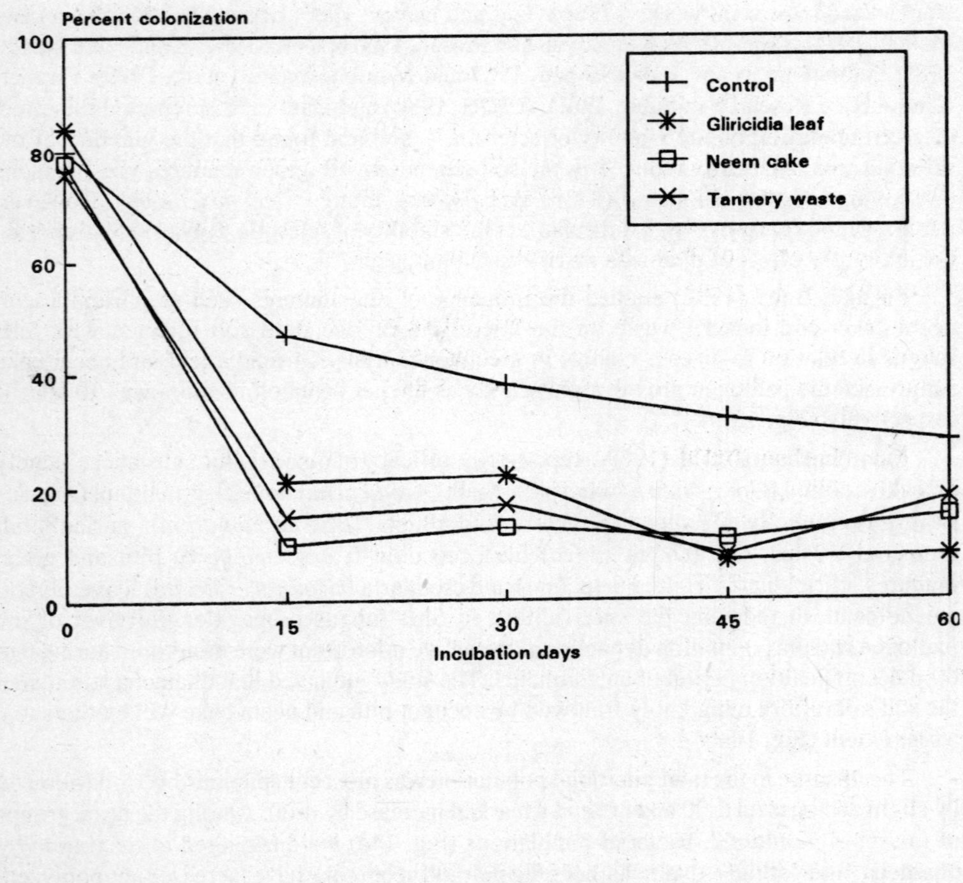

Fig. 13 : Effect of amendments on saprophytic survival of *R. solani* in soil (Baby, 1992).

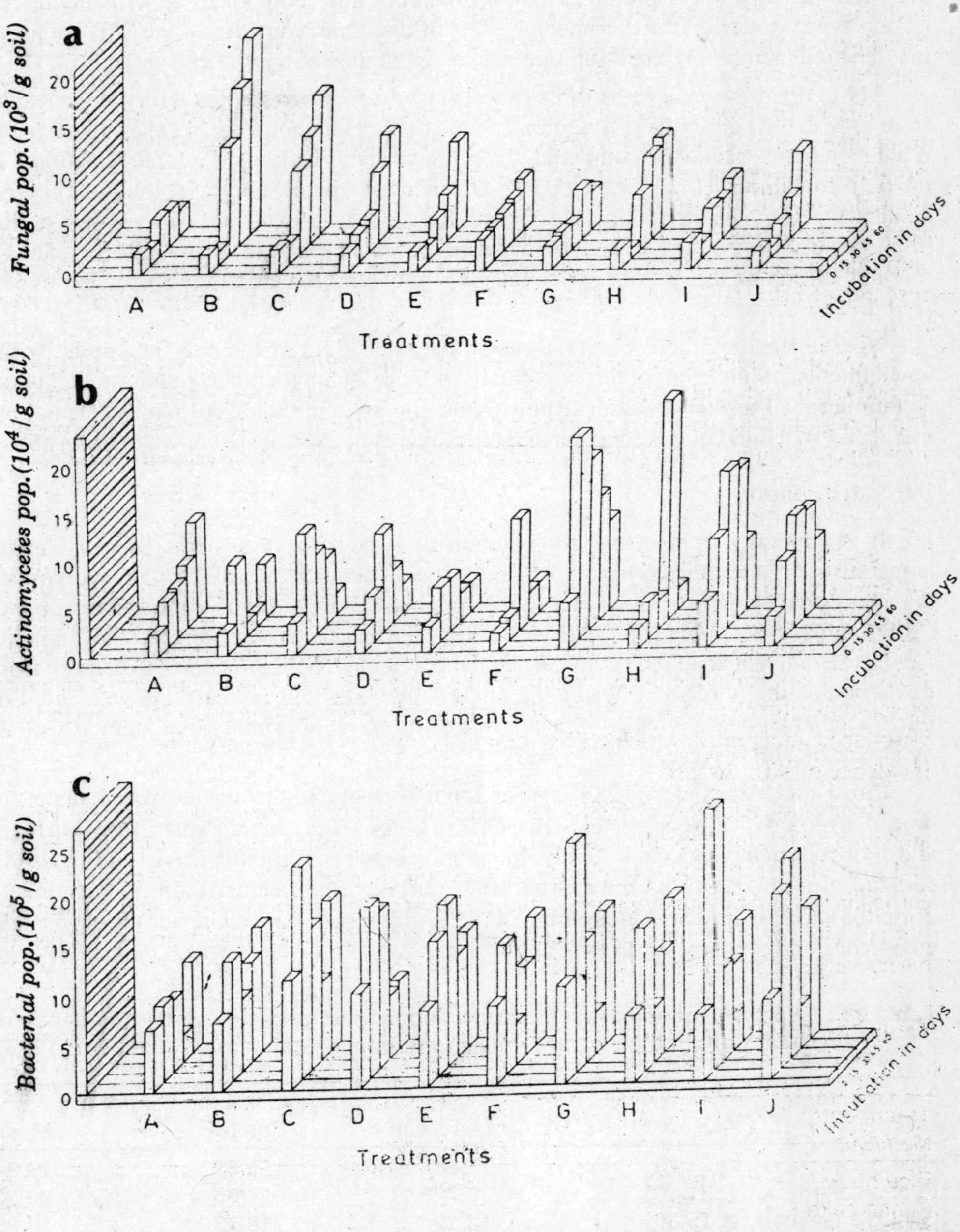

Fig. 14abc : Effect of amendments on soil microflora, (a) fungal population, (b) actinomycete population and (c) bacterial population. A=control, B=cotton seed cake, C=neem cake, D=groundnut cake, E=gingelly cake, F=illupei cake, G=coconut pith, H=rice husk, I=dhaincha leaf and J=crotalaria leaf (Manibhushanrao, et. al., 1989a).

in the number of pathogen propagules by inactivation either temporarily or permanently can be achieved. However, the detrimental effect of these amendments on the ShB pathogen may be mediated through soil saprophytes, or decomposition products on the host.

The major draw back with the use of soil amendments is irreproducibility of the results from time to time under field conditions. Some of the greenhouse/field experiments using organic manures cannot be compared, the physical status differ and seldom was found in the field conditions. However, in recent years increased interest in such biological control studies to eliminate or reduce the use of fungicides coupled with awareness of the importance of protecting the ecosystem and environment has been the impetus for renewed research efforts on the use of organic manures to control the ubiquitous pathogen like *R. solani*.

Notwithstanding these observations on the microflora of the rice field soils by the amendments, it should be emphasized that the primary claim for most organic amendments as antimicrobial substances is not disputed here. The sole aim should be to study the possible imbalance in natural microflora that organic manures in general are known to bring about.

Integrated control

It is generally opined that combination of two or more systems such as fungal antagonists, organic amendments, PAB, and systemic fungicides provide high level control than that could be achieved with each method alone. This would expose the soil saprophyte (ShB fungus) to a 'double/triple barrier' whereby the resistance would protect the fungicide from a tolerant isolate of the pathogen and the fungicide would protect the resistance from novel compatible virulence. While the organic manures will increase the residential or introduced antagonists which will again reduce the level of primary inoculum as well as incidence of ShB disease.

Rajan and Alexander (1988) earlier reported that ShB incidence was reduced in amended plots with gliricidia leaves, rice husk or lime as well as antagonist *T. viride* in main and split-plot treatments respectively. Interaction between gliricidia leaves and *T. viride*/carbendazim and rice husk/lime with carbendazim were significant. Among the treatments gliricidia leaves or lime amended plots recorded the lowest disease incidence in both panicle emergence and preharvest stages (Tables 23 and 24).

Table 23 : Effect of amendments, biocontrol agent and fungicide on the incidence of sheath blight disease at panicle emergence (Rajan and Alexander, 1989).

Treatments Main/Sub	*T. viride*	Carbendazim	Control	Mean
Rice husk	21.75	15.64	20.62	19.34
Punna cake	21.20	20.72	25.65	22.52
Gliricidia leaves	15.86	12.19	19.73	15.93
Lime	17.77	16.47	22.89	19.05
Gypsum	26.89	16.62	32.57	25.36
Control	39.16	21.05	47.36	35.86
Mean	23.77	17.12	28.14	

C.D. (P = 0.01) Main treatments 4.54 Sub-treatments 3.21 Interaction 7.86

Table 24 : Effect of amendments, biocontrol agent and fungicide on the incidence of sheath blight disease before harvest (Rajan and Alexander, 1989).

Treatments Main/Sub	*T. viride*	Carbendazim	Control	Mean
Rice husk	23.81	18.55	26.78	23.05
Punna cake	25.84	23.71	31.74	27.10
Gliricidia leaves	21.12	17.20	21.13	19.82
Lime	19.46	18.66	25.81	21.31
Gypsum	28.74	22.35	35.63	28.91
Control	42.41	26.94	48.32	39.22
Mean	26.90	21.24	31.57	
C.D. (P = 0.01)	Main treatments 4.25		Sub treatments 3.01	Interaction 7.37

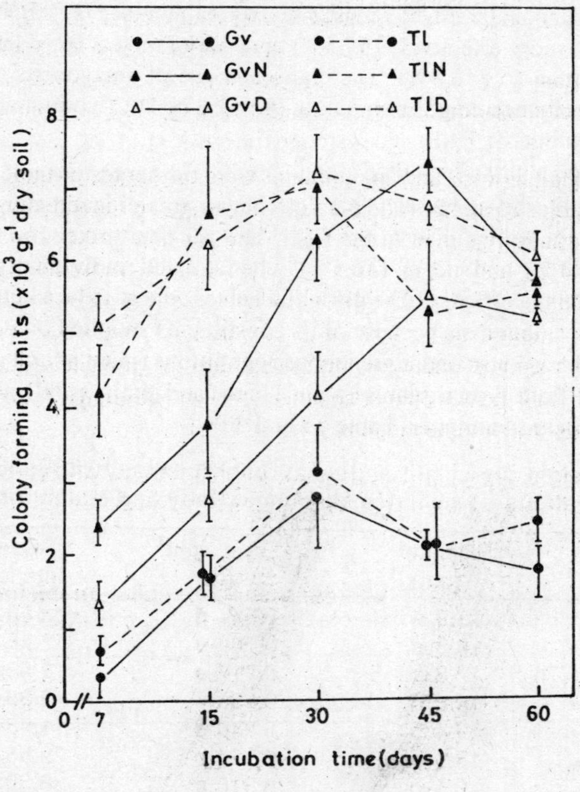

Fig. 15 : Survival of introduced antagonists in un-supplemented soil and in soil supplemented with organic materials (error bars indicate standard error). Gv. = *Gliocladium virens*, GvN= *G. virens*+neem cake, GvD=*G. virens* + gliricidia, T'I = *Trichoderma longibrachiatum*, TIN= *T. longibrachiatum* + neem cake, TID=T. *longibrachiatum* + gliricidia (Baby and Manibhushanrao, 1993).

Viji et al. (1991) reported that the ShB pathogen and the mycoparasites are different in their response to systemic fungicides. Antagonistic fungi such as *G. virens*, *T. harzianum* and *T. longibrachiatum* are tolerant to organophosphorus groups of fungicides while the ShB pathogen and the three antagonists are highly sensitive to benzimidazole compounds. These results indicate that an integrated approach along with a specific group of fungicides which are less toxic to residential antagonists appear to be promising.

Baby and Manibhushanrao (1993) studied the population density of antagonists *G. virens* (Gv) and *T. longibrachiatum* (T1) in amended soil and found a significant increase in the colony forming units (CFU) of these antagonists when they were introduced in soil with gliricidia leaf and neem cake (Fig. 15). The populations of Gv and T1, 60-d after incubation (DAI), were 5.67 x 10^3 and 6.08 x 10^3 (with gliricidia) compared with 1.76 x 10^3 (neem cake) and 2.46 x 10^3 in control (without amendment), respectively. Gliricidia leaf was found to be a better substrate for the population growth of antagonists than neem cake. This is indicative of their better survival in the amended soil. The pathogen's (*R. solani*) survival was considerably decreased by all treatments (Fig. 16). Neem cake suppressed the pathogen more effectively (57.5%) than gliricidia leaf (65%) and Gv was more effective (51.8%) than T1 (48.4%). The combined use of antagonists and amendments substantially reduced the pathogen population (40%) at 60 DAT compared with 80% in the control.

The efficacy of antagonists and amendments were further tested under greenhouse and field conditions (Tables 25 and 26) and their effect was greater in reducing the ShB disease under greenhouse conditions than in the field. The per cent protection was greater with gliricidia leaf (42.4%) and neem (40.5%) when treated individually, but combined treatments increased the efficacy. The gliricidia leaves appear to be a better substrate than neem cake for both antagonists because of its constituents supplied on its decomposition. Besides, the growth response under greenhouse conditions (plant height and dry weight of the seedlings) and field grown plants (plant height and grain yield) were significantly increased in combined treatments (Table 25 and 26).

Table 25 : Plant height, dry weight, and sheath blight intensity with various amendments in greenhouse-grown rice seedlings (Baby and Manibhushanrao, 1993).

Amendment	Height[1] (cm)	Dry weight[1] (g)	Disease intensity[1]	Per cent protection[2]
Control (C)	13.9 e	0.541 g	26.20 a	-
G. virens (Gv)	16.3 d	0.708 e	12.46 bc	52.4
T. longibrachiatum (TI)	17.2 cd	0.717 e	13.52 b	48.4
Gliricidia (D)	16.3 d	0.650 f	10.38 cde	60.4
Neem (N)	19.4 b	0.746 d	11.43 bcd	56.4
Gv + D	18.1 c	0.764 d	9.55 de	63.6
TI + D	19.1 b	0.810 c	8.50 e	67.6
Gv + N	22.6 a	0.900 a	8.09 e	69.1
TI + N	22.8 a	0.847 b	8.19 e	68.7

Values followed by the same letters are not significantly different (P = 0.01) according to Duncan's Multiple Range Test

[1] Measured 30 DAS

[2] Per cent protection = $\frac{\text{Disease in control - disease in treatment}}{\text{Disease in control}} \times 100$

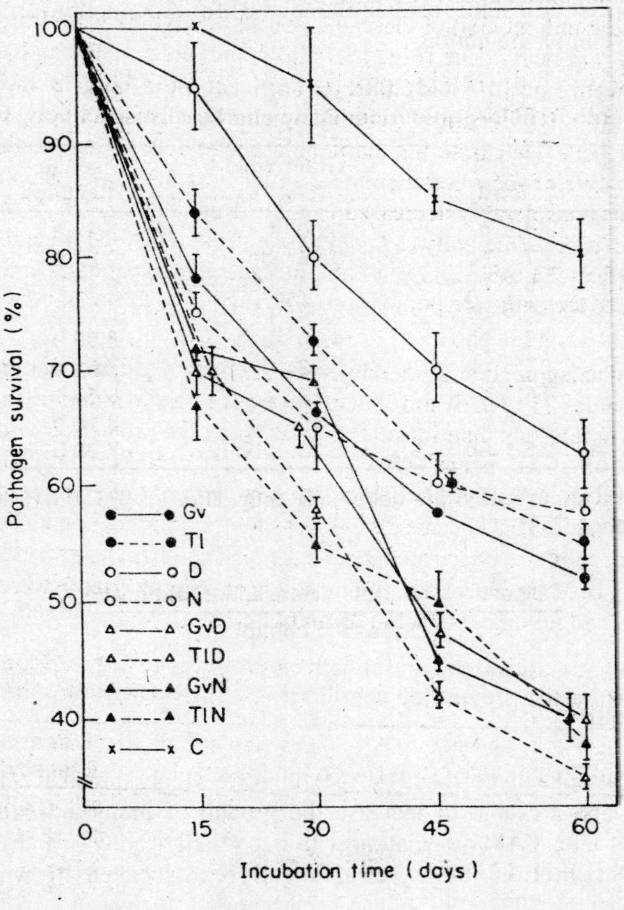

Fig. 16 : Survival of the pathogen *R. solani* in various soil treatments (error bars indicate standard error). C=control, N=neem cake, D=gliricidia, Gv. + *Gliocladium virens*, GvN = *G. virens* + neem cake, GvD = *G. virens* + gliricidia, TI = *Trichoderma longibrachiatum*, TIN = *T. longibrachiatum* + neem cake, TID = *T. longibrachiatum* + gliricidia (Baby and Manibhushanrao, 1993).

In the integrated control, pectic enzymes produced by the antagonists might have accelerated the degradation of the organic manure amendments and release of nutrients leading to enhanced plant growth and yield and may be the adjunct mechanism of action of the combined treatments. Thus, integrated disease management (IDM) of rice ShB might be both economically and environmentally viable and therefore the prospect of adoption in the control of this dreadful soil-borne pathogen. IDM should be given top priority in ShB management as it combines low input technology, utilizing natural resources (agro-industrial waste products and residues of plant and animal life). Admittedly, further research is needed for effective implementation.

Table 26 : Plant height, grain yield, and sheath blight intensity with various amendments in field-grown rice (Baby and Manibhushanrao, 1993)

Amendment	Height[1] (cm)	Grain Yield (t ha^{-1})	Disease intensity[1]	Per cent protection[2]
Control (C)	44.8 c	3.00 f	15.11 a	-
G. virens (Gv)	47.4 bc	3.94 de	9.27 b	38.6
T. longibrachiatum (Tl)	48.8 abc	4.44 cde	9.47 b	37.3
Gliricidia (D)	46.5 bc	3.62 ef	8.71 b	42.4
Neem (N)	49.8 abc	4.60 bcd	8.99 b	40.5
Gv + D	55.5 ab	5.92 a	7.64 b	49.4
Tl + D	56.7 a	5.73 a	6.66 b	55.9
Gv + N	57.8 a	5.40 ab	8.15 b	46.1
Tl + N	55.1 ab	5.33 abc	7.29 b	51.8

Values followed by the same letters are not significantly different (P = 0.01) according to Duncan's Multiple Range Test

[1] At booting stage

$$^{2}\text{Per cent protection} = \frac{\text{Disease in control - disease in treatment}}{\text{Disease in control}} \times 100$$

Mycorrhizal association

The vesicular arbuscular mycorrhizal (VAM) fungi are widely spread in cultivated soil (Gerdemann, 1968) and are able to increase the growth of many agricultural crops (Campbell, 1989). Of late, VAM association with rice (Gangopadhyay and Das, 1988); Ammani et al. 1985; Brown et al. 1988) as well as fungal types associated (Brown et al. 1988, Baby and Manibhushanrao 1992; 1995) have been reported. Increased growth response (Sanni, 1976 and Khan et al., 1988) and reduced disease incidence (Gangopadhyay and Das, 1988) due to rice plant - VAM fungal association is known. Further, Baby (1992), Baby and Manibhushanrao (1992) did extensive survey of the VAM fungi in the rice field soils of Kerala State (South India) where rice is grown under diverse cultural conditions. The presence of VAM fungal spores in diverse soil conditions has been established. The investigation indicated that *Glomus aggregatum, G. ambisporum, G. fulvum, G. fasciculatum, G. occultum* as dominant and *G. mosseae, G. microaggregatum, Acaulospora* sp., *Gigaspora gigantea* and *G. candida* as common forms in rice fields. A few among them are given in Plate 16.

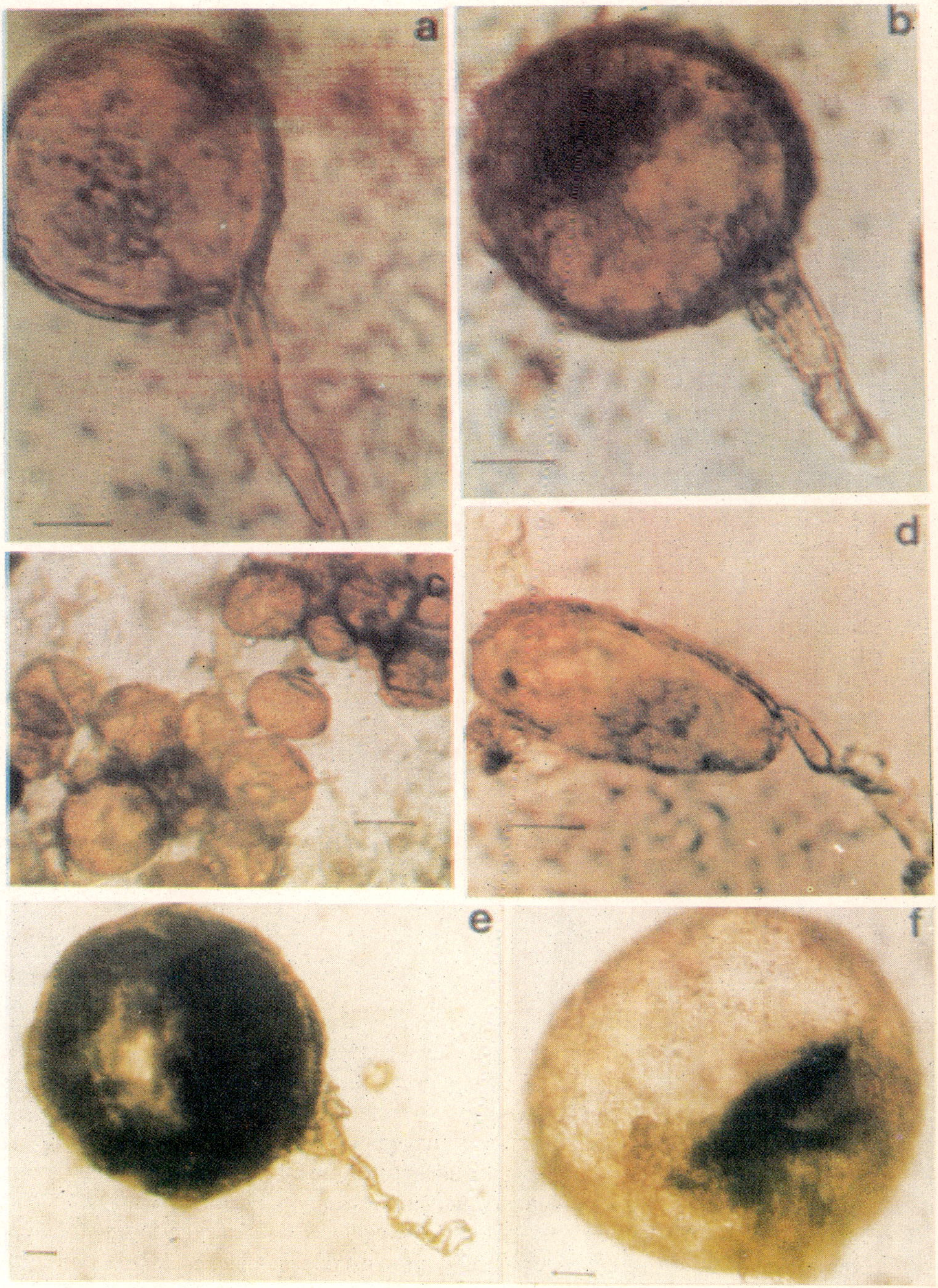

Plate 16 ABCDEF : VAM fungal spores isolated from rice field soils (Bar = 20μm), (A) *Glomus faciculatum*, (B) *G. mosseae*, (C) *G. aggregatum*, (D) *G. fulvum*, (E) *Gigaspora candida* and (F) *G. gigantea* (Baby, 1987).

Baby and Manibhushanrao (1995) made a periodical assessment of mycorrhizal association of rice roots and estimated the spore density under greenhouse/field conditions. Twelve organic amendments (Table 27) of which five oil seed cakes, four green manures and three agro-industrial wastes at 0.5% level under greenhouse (15 x 20 cm earthen pots) as well as field ($1 m^2$ micro plots) trials were incorporated in the soil. This study revealed that an increase in the spore density was observed from 30-d onwards in all amendments in greenhouse grown plants except the agro-industrial wastes. Under field condition, the spore number increased up to 45-d, after which there was not much of variation till booting stage. However, in the field condition an increase in spore number was observed from booting to grain filling stage of the plants (Table 28).

Table 27: Influence of organic amendments on VAM fungal spore density and infection of rice roots in greenhouse grown plants (Baby and Manibhushanrao 1993)

	Age of the plant (days)											
	5		10		20		30		40		50	
Amendment	Spn*	PI	Spn	PI	Spn	PI	Spn	PI	Spn	PI	Spn	PI
Control (unamended soil)	58	28.7 (1)	75	32.0 (1-2)	112	31.3 (1)	165	40.7 (1-2)	190	63.3 (1-2)	205	78.0 (2-3)
Marotti cake	60	22.0 (1-2)	56	39.3 (1-3)	92	53.3 (1-2)	180	60.7 (1-2)	213	66.7 (3)	280	89.3 (1-2)
Neem cake	52	36.7 (2)	68	44.6 (3)	66	59.3 (2-3)	138	74.0 (2-3)	240	82.0 (3)	315	94.0 (2-3)
Punna cake	50	26.0 (1)	54	38.7 (1)	102	45.3 (1-2)	96	52.7 (1-2)	186	61.3 (1-2)	232	80.0 (1-2)
Rubber seed cake	62	21.3 (1)	70	32.7 (1-3)	106	49.3 (1-2)	176	52.0 (1-2)	224	65.3 (1-2)	219	70.0 (1-2)
Eupatorium leaf	36	6.7 (1)	48	12.0 (1)	72	29.3 (1)	105	46.7 (2-3)	158	57.3 (1-2)	244	68.0 (1-2)
Gliricidia leaf	28	7.3 (1)	56	22.7 (1-2)	54	32.7 (1-2)	125	44.0 (2)	248	60.7 (1-2)	286	81.3 (2-3)
Cassia leaf	22	27.3 (2)	42	48.0 (1-2)	60	58.7 (1-2)	128	53.3 (1-2)	240	65.3 (2-3)	230	86.7 (3)
Neem leaf	25	10.0 (1)	46	27.3 (1-2)	68	50.0 (1)	98	50.7 (1-2)	210	58.0 (1-2)	215	70.0 (2-3)
Bonemeal	65	11.3 (1)	78	21.3 (1-2)	136	26.7 (1)	120	46.0 (1-2)	160	60.0 (2-3)	186	68.7 (1-2)
Coconut pith	60	19.3 (1)	83	25.3 (2-3)	148	31.3 (1-2)	156	51.3 (3)	190	78.7 (2-3)	252	97.3 (2-3)
Poultry manure	57	16.0 (1-2)	53	26.7 (1)	140	32.7 (1-2)	201	38.0 (2-3)	258	68.7 (2)	266	82.7 (1-2)
Saw dust	59	8.7 (1)	70	16.0 (1)	140	18.7 (1-2)	132	43.3 (3)	182	61.3 (2-3)	215	73.3 (2-3)

* Initial spore number (52 spores/50g dry soil)
Spn - Spore number (/50g dry soil)
PI - Per cent root infection
Values in parenthesis indicate intensity of infection (rating 1-3)

Table 28 : Mycorrhizal infection and response of rice cv. TKM 9 to Sheath blight incidence at different stages of growth (Baby and Manibhushanrao, 1995)

	Booting stage				Grain filling stage			
	Healthy	Infected	T-value	Sig.	Healthy	Infected	T-value	Sig.
Spore number*	203 (7.2)	186 (6.6)	3.02	0.05	400 (8.9)	340 (4.4)	10.47	0.01
Per cent infection	60 (1.2)	45 (1)	16.63	0.01	74 (5.2)	15 (1.5)	6.40	0.01
Number of vesicles@	35 (2.7)	18 (2.7)	7.71	0.01	56 (2)	32 (2.7)	12.37	0.01
Infection intensity	2-3	1-2			2-3	1-2		

*Number of spores/100g dry soil; @ Number of vesicles/root segment
Values under healthy and infected treatments are mean of triplicate
Value in paranthesis indicate SD of triplicate; Sig.: Significant level

In general the per cent of infection (microscopic observations of 100 root segments) and intensity of infection (Kormanik et al., 1980) were low up to 45-d and increased thereafter (60-d) and reached maximum in the grain filling stage. Further, the assessment of VAM colonization in healthy and *R. solani* infected rice plants revealed that spore density in root zone as well as per cent infection were more in healthy plants than in diseased plants both at booting and grain filling stages (Table 29). The intensity of infection was 2-3 rating in healthy plants while it was less (1-2 rating) in the diseased plants. The per cent as well as the intensity of VAM infection and number formed in the roots and the spore density were high in healthy compared to diseased plants. Better nutrition might have enchanced vigour and defence oriented physiological alterations which might be the chief mechanism in VAM associated plants in countering the potential pathogens. These results also suggested that VAM fungi resist the attack of *R. solani* and the potential organic amendments do enhance the activity. Though the mechanism(s) involved in disease resistance is complicated, any doubt on the involvement of VAM fungi as a part of the complex process cannot be ignored.

Saprophytic phase in different soils

Application of systemic fungicides, organic soil amendments, antagonists, plant growth promoting rhizobacteria (PGPR) as well as plant associated bacteria (PAB) on the saprophytic survival of ShB pathogen *R. solani* is well known. However, Venkatraman et al. (1995) studied the saprophytic phase of the ShB pathogen *R. solani* in a disease suppressive (vertisol of Hyderabad, Andhra Pradesh) and two conducive soils (sandy clay of Pattambi and silty loam of Moncompu, Kerala). In the conducive soil of Moncompu, highly significant growth was recorded while in Pattambi the fungus growth was on par with the suppressive vertisol of Hyderabad (Table 30). However, in all the three soils the pathogen growth was negligible. This study showed the plausibility of locating ectotropic saprophytic growth of *R. solani* as such in the soils which harbour the potential antagonists to *R. solani*.

Table 29 : Influence of organic amendments on VAM fungal spore density and infection of rice root in field grown plants (Baby and Manibhushanrao 1993)

Amendment	Age of the plant (days)											
	15		30		45		60		Booting		Panicle filling	
	Spn*	PI	Spn	PI	Spn	PI	Spn	PI	Spn	PI	Spn	PI
Control (unamended soil)	249	41.7 (1)	407	25.0 (1)	380	8.3 (1)	253	33.3 (1)	323	66.7 (1-2)	370	79.2 (1-3)
Marotti cake	170	8.3 (1)	383	33.3 (1)	452	16.7 (1)	480	41.7 (1-2)	396	79.2 (1-3)	363	83.3 (2-3)
Neem cake	118	58.3 (1)	247	37.5 (1)	713	16.7 (1)	546	43.3 (2)	560	79.2 (2-3)	385	91.7 (2-3)
Punna cake	133	33.3 (1)	235	24.0 (1)	734	18.8 (1)	526	48.8 (1-2)	549	62.5 (2-3)	368	87.5 (1-3)
Rubber seed cake	162	25.0 (1)	172	24.0 (1)	308	33.3 (1-2)	370	45.8 (1-2)	425	55.8 (1-2)	394	83.3 (1-2)
Eupatorium leaf	92	29.2 (1)	211	37.5 (1)	723	29.2 (1)	566	39.7 (1-2)	680	62.5 (1-3)	805	83.3 (1-3)
Gliricidia leaf	89	33.3 (1)	266	25.5 (1)	550	33.3 (1)	327	37.5 (1-2)	446	62.5 (2-3)	626	66.7 (1-3)
Cassia leaf	108	58.3 (1-2)	249	12.5 (1)	530	16.7 (1)	466	33.3 (1-2)	498	95.8 (2-3)	590	79.2 (1-3)
Neem leaf	95	50.0 (1)	131	29.2 (1)	578	16.7 (1)	425	37.5 (1-2)	515	87.5 (2-3)	698	87.5 (1-3)
Bonemeal	250	25.0 (1)	234	45.8 (1-2)	584	37.5 (1-2)	551	29.1 (1-3)	379	95.8 (1-3)	570	87.0 (1-2)
Coconut pith	307	25.6 (1)	396	50.0 (1)	597	45.8 (1-2)	310	28.7 (1-2)	387	62.5 (1-3)	448	75.0 (1-3)
Poultry manure	292	37.5 (1)	507	25.7 (1)	517	8.3 (1)	331	12.5 (1-2)	482	61.1 (3)	632	91.7 (1-3)
Saw dust	290	20.8 (1)	228	41.7 (1)	472	33.3 (1)	329	45.8 (1-2)	353	83.3 (1-3)	496	87.5 (1-3)

*Initial spore number of the field soil (115 spores/100g dry soil)
Spn - spore number (/100g dry soil)
PI - per cent root infection
Values in parenthesis indicates intensity of infection (rating 1-3)

Table 30 : Saprophytic growth of *Rhizoctonia solani* in different soils (Venkatraman et al., 1995)

Soil	pH	Growth of *R. solani* (mm) after days of incubation		
		7	14	21
Sandy loam (Pattambi)	5.5	18	23	23
Silty clay (Moncompu)	4.9	35	36	37
Vertisol (Hyderabad)	7.9	18	21	21

Treatment-CD (P = 0.05)
7-d 7.34

Models for biocontrol

Considering the biocontrol models for soil borne disease, unlike the parameters involved with epidemiology of foliar pathogen like rice blast disease, here the model works out for a fixed space i.e., rhizosphere soil and the disease should react in a describable and predictable ways in the infection process in a relative stable environment and ecosystem. Rouse and Baker (1978) attempted to model a biocontrol system, while, Gilligan (1983) developed models based on soil borne pathogens. Baker (1990) has modelled the effects of cellulose and chitin amendments in biocontrol to reduce the soil borne pathogen *R. solani* from propagules in the spermosphere of radish seeds by the following equation:

$$A = \int_{x = I_0}^{x = I_1} \{(b_1 - b_1)\, x + (b_0 - b_0)\}\, dx$$

where, A equals the change in the area under the curves due to control, b_0 and b_1 are the regression coefficients for the raw soil curve, b_1 and b_0 are the regression coefficients for the second treatment curve and I_1 and I_0 are the inoculum density limits for integration. In conclusion, the results suggest that cellulose amendment was more efficient than that of chitin to reduce the *R. solani* propagules.

Johnson (1994) proposed a model that defines parameters that affect the efficacy of biocontrol agents on reducing the impact of plant pathogen population. The model was developed exclusively based on analogy to the most common dose-response relationship observed for plant pathogen and host interaction with plant parts (tissue). The pathogen dose (x)—plant disease response (y) can be modelled as follows:

$$y = 1\text{-exp}\ (-kx)$$

where x is the inoculum density of the pathogen, k is a constant that governs the efficiency of the inoculum, and y is the proportion of plant pathogen disease. The above model assumes 100% plant or plant tissue is susceptible to infection, but normally some plant parts may not be affected by infection. An asymptote, L can be introduced to a dose-response model to exclude unavailable tissue or plant part and the above model becomes as $y = L[1\text{-exp}(-kx)]$.

A model to know the effect of a biocontrol agent (z) on a plant-pathogen population(x) is as follows:

$$x_i/x = A\ [1\text{-exp}\ (-cz)],$$

where, z is the inoculum density of the biocontrol agent, c is a constant that governs the efficiency of the biocontrol agent, x_i/x is the proportion of inoculum units of the plant biocontrol population not effected by the biocontrol agent and A represents the maximum proportion of a pathogen population that can be ineffective due to a biocontrol agent. The proportion of disease (y) or the amount of disease in response to the density of a biocontrol agent (z) at different levels of the pathogen (x) is described as:

$$y = 1\text{-exp}\ [-kx\ \text{exp}(-cz)]$$

The above model holds good only when all the host tissue is susceptible and all the pathogenic inoculum is completely affected by a biocontrol agent. The term exp(-cz) in the above model defines the proportion of pathogenic inoculum that remains effective (i.e., 1-x_t/x). The model can be represented to include the effect of some unavailable plant tissue on disease control due to biocontrol agent. Thus, the model becomes:

$$y=1-\exp(-kx\{1-A[1-\exp(-cz)]\})$$

The sensitivity of the above model in reducing the amount of disease due to biocontrol agent will depend on two efficiency parameters (k and c), two inoculum density variables (k and z) and an asymptote parameter (A).

An empirical model building was studied by Hadar et al. (1979) using antagonist, *T. harzianum*, grown on wheat bran and incorporated into soil against damping-off of eggplant caused by *R. solani* is given in Fig. 17.

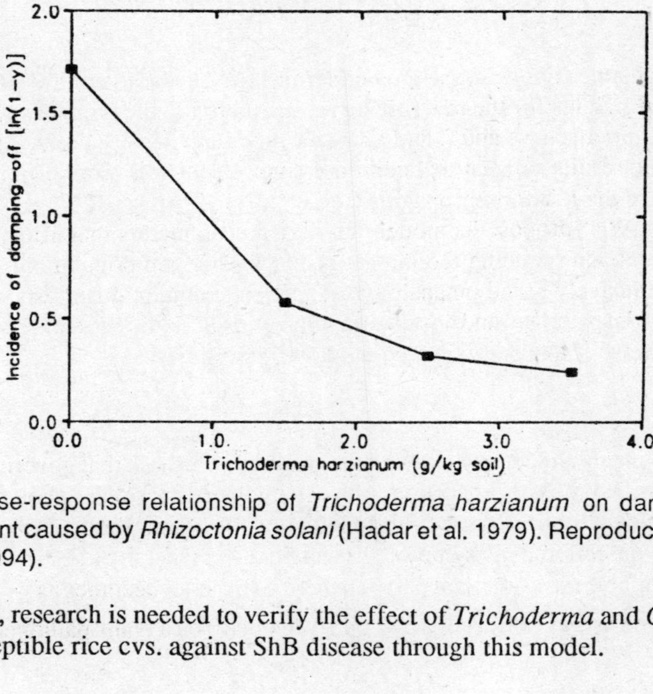

Fig. 17 : Dose-response relationship of *Trichoderma harzianum* on damping off of egg plant caused by *Rhizoctonia solani* (Hadar et al. 1979). Reproduced from Johnson (1994).

Further, research is needed to verify the effect of *Trichoderma* and *Gliocladium* spp. on the susceptible rice cvs. against ShB disease through this model.

Botanical pesticides

Tewari and Dath (1984) reported that leaf extracts of henna or tree mignonette (*Lawsonia inermis*), betel leaf (*Piper betle*) and bale tree (*Aegle marmelos*) are toxic to mycelia and sclerotia of ShB fungus. Extracts of these plants are less toxic to rice plants. Shukla et al. (1990) reported a few essential oils of *Citrus sinensis, Eucalyptus globulus, Ocimum canum* and *Pinus roxburghii* which are toxic to ShB fungus indicating the presence of antifungal compounds. Further, crystalline mimosine at 4 mM and the crude extract of mimosine from the seeds of *Leucaena leucocephala* at 1:10 dilution completely inhibited the growth of *R. solani* indicating the possibility of exploiting mimosine as biocide (Radha, 1991). Tewari and Mandakini (1991) further studied the antifungal activities of leaf extracts of *P. betle*, *O. sanctum*, *Nictanthes arbor-tristis* and *Citrus lemon* against *R. solani* and two other fungal pathogens on rice. However, the extracts of *P. betle* and *O. sanctum* were very effective in reducing the growth completely *in vitro* while checking the spread of the ShB disease *in vivo*. They suggested that these two extracts are the best source as biopesticides in combating the principal fungal diseases of rice.

Since pesticides of plant origin are systemic in nature and easily biodegradable, clearly, there is a vast scope to identify plants possessing such antifungal substances toxic to ShB fungus. More critical evaluation of botanical pesticides/biocides with field grown plants is required.

Challenges for the Future

Preamble

Three decades ago, the last international symposium on "*Rhizoctonia solani*-Biology and pathology" was hosted by the American Phytopathological Society (APS) in Florida, USA. Since then there have been major advances in our understanding of the *Rhizoctonia* spp. causing economically significant diseases world wide on most important crop plants including rice. New techniques in cell and molecular biology and accurate prediction of yield losses through simulation models have generated much data and new perspectives. Development of environmentally acceptable biocontrol strategy through the suppression of plant pathogens provide a reliable and safe alternative to much pesticide abuse. The forthcoming international symposium on *Rhizoctonia* (ISR) in 1995 organized by ISR-Rhizoctonia working group at Noordwijkerhout, the Netherlands, is therefore timely and will enlighten us on the update of advances in *Rhizoctonia* research. However, this monographic study is aimed exclusively on the rice ShB disease with ample coverage on current perspectives and challenges.

Among the fungal pathogens, rice ShB fungus (*R. solani*) is considered as one of the major production constraints for rice in South East Asia where one third of the world's five billion people live and depend on rice as major staple food. This is largely with the introduction of IR-type semi-dwarfs, and with the advancement of appropriate technology for increasing productivity. Intensive cultural practices have lead to the failure of the crop against ShB disease.

Two major challenges facing against the rice ShB disease control are, firstly, to manage the rhizosphere pathogen propagules in the present intensive areas of cultivation and to minimize the crop yield and secondly, to prevent the spread of ShB disease that are at present geographically limited. Success in meeting these challenges will depend upon the clear understanding of genetic and molecular biology of the ShB fungus. In this concluding section, any gaps in our knowledge on ShB control have also been identified. The priority areas of ShB research are summarized and further intensive research on genetic analysis for strain improvement and its potential prospects envisaged is discussed. Further, the number of publications appeared on the rice ShB disease and *R. solani* indicated from rice database, IRRI for the years 1980-94 shows that a total of 166 publication appeared during 1980-84, while there has been a spurt on the contributions (244) during 1985-89. However, there has been a decline (173) in the number of contributions during 1990-94. This trend is the same with both rice ShB disease as well as on the pathogen *R. solani* (Fig. 18).

Economic loss

Relationships between disease intensity/severity and loss of crop yield need to be established at various levels of crop management using the traditional cvs. with high levels

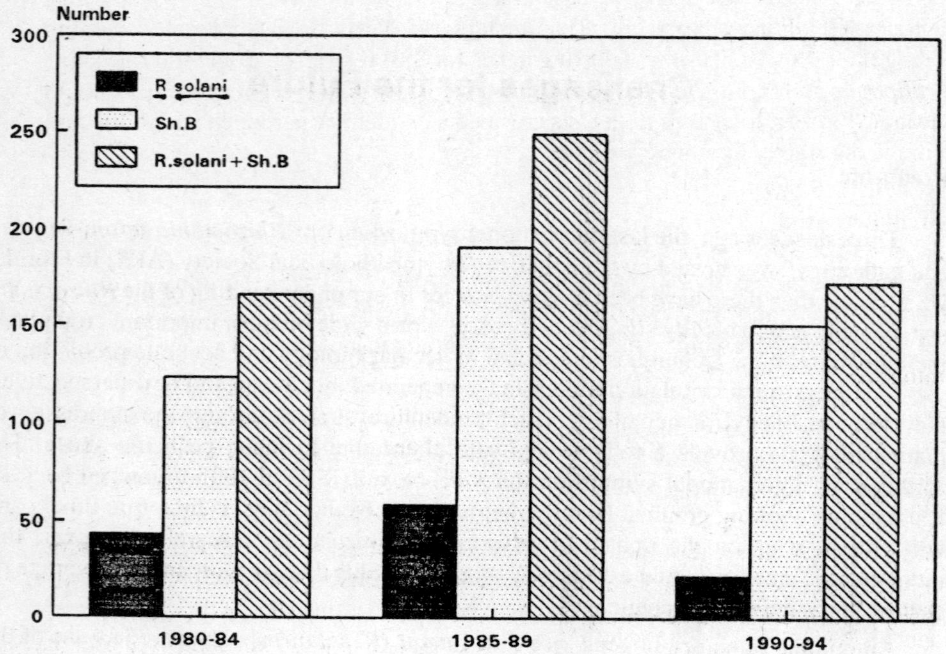

Fig. 18 : Number of publications on rice ShB disease and *Rhizoctonia solani* during 1980-94 (Courtesy : IRRI Bibliographer).

of resistance. Realistic estimates of annual crop losses should be determined periodically at national/international level using optimum N-input. These estimates may also be used to develop computer simulation models for accurate prediction of ShB disease.

Cultural practices

These include (i) destroying the suspected collateral hosts near the vicinity of rice fields as well as burning the stubbles, (ii) deep ploughing of the fields in summer to expose the sclerotia to sunlight, (iii) seed treatment with a systemic fungicide/spores potential with suitable with wetting agent or seed bacterization, (iv) avoidance of thick sowing seeds especially dapong nurseries, (v) preventing close planting and providing wider spacing of transplanted seedlings to reduce high humidity in the plant ecosystem, (vi) avoiding excess N-fertilizers on high yielding cvs. and (vii) judicious application of greenmanures especially gliricidia leaves before puddling the soil to reduce the pathogen propagules.

Host resistance

Knowledge of the inherent level of resistance in rice cvs. to ShB disease is far from adequate. There is a great need for more work to identify resistance genes from donor parents. The relevance on the effective screening to ShB in regions where sophisticated, fast and accurate methods using biochemical parameters are to be used as a large scale screening of breeding lines. Possibility of transferring genes for ShB resistance from wild rices viz., *Oryza rufipogon* and *O. barthii* which are known to possess high level of resistance should be envisaged. Further, long term field tests are needed to identify pathogen variability and to maximize the stable/durable resistance.

Biocontrol potential

Prospects for the management of ShB through biocontrol especially integration of potential antagonists, and organic manures and to some extent with VAM fungi was focused in mid 80's as one of the research priorities. The potential prospects envisaged in biocontrol are summarized.

Application of wheat bran saw dust (WBSD) preparation of *Trichoderma* spp. has been useful to control several soil-borne root infecting pathogens under field conditions. This study should be extended on a priority basis to rice ShB control. A suitable, cheap and freely available substrate can be recommended.

Plant growth promoting rhizobacteria (PGPR) and plant associated bacteria (PAB) are known to colonize or compete or suppress the deleterious microorganisms as well as soil-borne pathogens at the root surface. Such potent bacterial strains should be used as seed inoculants and tested under field conditions to control rice ShB disease. Further, growth promoting effort of the plants by biocontrol agents can be attributed as a 'biofertilizer' effect. Detailed biochemical investigations are necessary to work out the growth promoting substance(s) produced by the antagonists. This dual advantage by the antagonists should be exploited in the management of rice ShB disease. Genetic manipulation of the antagonists through mutation, recombination, gene transfer or protoplasmic fusion to improve their antibiotic production is quite essential and needs immediate attention.

With the advent of recombinant DNA technology, it becomes thus possible to isolate new genes to our advantage and they should be incorporated in DNA sequence that regulate their spatial expression within the transgenic plant with one or more additional genes. Oflate, molecular biologists from IRRI, Philippines have successfully transfered genes for insect resistance by engineering rice with genes from bacteria and soyabean. Genes from the potent antagonists with enhanced antibiotic production can also be transferred to rice against ShB pathogen. Our future effort in managing rice ShB disease should be through biotechnology-a novel means of mobilizing and incorparating alien genes to rice pool through genetic engineering.

Antagonists namely fluorescent pseudomonads and *Bacillus subtilis* used for seed bacterization, increased seed germination, seedling vigour and biomass under greenhouse condition in several crop plants and has a potential to work on large scale field trials. Integrated approach with promising antagonists like *T. longibrachiatum* and *G. virens* reduced the pathogen propagules as well as sclerotial production of *R. solani* and *S. rolfsii* in soil is well known and this study under integrated disease management (IDM) should be fully exploited at national/international levels through coordinated field trials.

Liquid fermentation technology is used for the large scale production of metabolites of biocontrol agents utilizing the commercially viable, inexpensive substrates for the augmentation of seed treatment as they are considered superior with long term protection than the conventional fungicidal seed protectants used by many workers.

Registration of biopesticides of *Trichoderma* preparations in the developed countries is well known. Recently, the commercial preparation of *G. virens* was released in USA for the biological control of pathogens on vegetables and ornamentals. Biopesticides, compared with classical and potent pesticides are safe to handle, and environment friendly without any adverse effects. Thus, they open up new avenues for research and development.

The VAM fungi, of late, used as biofertilizers as well as biocontrol agents with suitable organic substrate(s), is possible to increase their association with rice plants especially under upland condition, where rice is directly seeded under dry conditions and ShB incidence is not uncommon. It is also possible that even under irrigated condition application of VAM as biofertilizers is a distinct possibility since it is now known that VAM fungi are associated with many plants in aquatic habitats. Incidentally, the area planted in upland rice is so large (about 19 million ha world wide to 37 million ha of irrigated lowland rice cultivation) that any marginal increase in yield would substantially influence the total rice production. This will ultimately boost the yield and an important component-the global agricultural economics.

References

Abeygunawardena, D.V.W. and E.M.B. Desilva. 1966. Chemotherapeutic investigations on sheath blight of rice. *Trop. Agric.* **120** : 149-158.

Adams, P.B. 1990. The potential mycoparasites for biological control of plant diseases. *A. Rev. Phytopath.* **28**: 59-72.

Ahn, S.W., R.C. Dela, B.L. Pena Candole and T.W. Mew. 1986. A new scale for rice sheath blight (ShB) disease assessment. *Intl Rice Res. Newsl.* **11** : 11.

Ahn, S.W. and T.W. Mew. 1986. Relation between sheath blight (ShB) and yield. *Intl. Rice Res. Newsl.* 11: 21-22.

Akai, S., H. Ogura and T. Sato, 1960. Studies on *Pellicularia filamentosa* (Pat.) Rogers. I. On the relation between the pathogenicity and some characters on culture media. *Ann. Phytopathol. Soc. Japan*, **25** : 125-130.

Aluko, M.O. and T.F. Hering. 1970. The mechanism associated with the antagonistic relationship between *Corticium solani* and *Gliocladium virens. Trans. Br. Mycol. Soc.* **55** : 173-199.

Amante, A.D., R.D. Pena, L.A. Sitch, L. Leung and T.W. Mew, 1990. Sheath blight (ShB) resistance in wild rices. *Int. Rice Newsl.*, **15** : 5.

Amaral, R.E., De.M., E. Issa, D.M. DeSouza, V.M.A. Malvolta, I.C. Leite and L.M.De. Jesus. 1979. Studies on sheath blight of rice (*Oryza sativa* L.). *Arguivos do Instituto Biologico,* **46** : 55-61.

Amin, K.S. 1975. An important method for evaluation of rice sheath blight *Phytopathology*, 65 : 214-215.

Anderson, N.A. 1982. The genetics and pathology of *Rhizoctonia solani. A. Rev. Phytopathol.*, **20** : 329-347.

Anderson-Prouty, A.J. and P. Albersheim. 1975. Host pathogen interaction. VIII. Isolation of a pathogen-synthesized fraction rich in glucan that evicts a defense response in the pathogen's host. *Pl. Physiol.*, **56** : 286-291.

Aoki, H., T. Sassa and T. Tamura. 1963. Phytotoxic metabolites of *Rhizoctonia solani. Nature, Lond.*, **200** : 575.

Araki, F. 1985. Monocut (Flutolanil), a new systemic fungicide. *Japan Pestic. Inf.*, **47** : 23-25.

Arunyanart, P., A. Surin and W. Rojanahasadin. 1984. Rice yield loss to sheath blight (ShB). *Intl. Rice. Res. Newsl.*, **9** : 10.

Atkins, J.G. 1974. *Rice diseases of the Americans. A Rev. Lit. US Dep. Agric. Hdbk.*, 448, pp. 106.

Awoderu, V.A. 1974. *Rice diseases in Nigeria.* PANS, **20** : 416-424.

Baby, U.I. 1992. Studies on the control of rice sheath disease blight with organic soil amendments. Ph.D. Thesis. Univ. of Madras, Madras. p. 128.

Baby, U.I. and K. Manibhushanrao. 1992. Vesicular-arbuscular mycorrhizal fungi of rice field soils. *Acta. Bot. Indica*, **20** : 10-15.

Baby, U.I. and K. Manibhushanrao. 1993. Control of rice sheath blight through the integration of fungal antagonists and organic amendments. *Trop. Agric.*, **70** : 240-244.

Babay, U.I and K. Manibhushanrao. 1995. Influence of organic amendments on vesicular-arbuscular mycorrhizal fungi in relation to rice sheath blight disease. *Mycorrhiza*, : (In press).

Baker, K.F. 1970. Types of *Rhizoctonia* diseases and their occurrence. In : *Rhizoctonia solani Biology* and *Pathology*, (ed.) J.R. Parmeter, Jr. Univ. of Calif. Press, Berkley, Los Angeles and London, pp. 125-148.

Baker, K.F. 1987. Evolving concepts of biological control of plant pathogens. *A. Rev. Phytopath.*, **25** : 67-85.

Baker, R. 1990. An overview of current and future strategies and models for biological control. In: *Biological control of soilborne plant pathogens.* D. Homby (ed.), CAB International, Lond. pp. 375-388.

Bandong, J.M., F.L. Nuque and E.V. Castro. 1975. Varietal resistance to sheath blight. Int. Rice Res. Inst. (IRRI), Saturday Seminar papers, Box 933, Manila, Philippines (mimeographed).

Basu, A. and P.K. Sengupta, 1992. Loss in yield and seed infection in promising genotypes of rice (*Oryzae sativa*) due to sheath blight disease caused by *Rhizoctonia solani*. *Indian J. Agric. Sci.*, **62** : 570-571.

Bhaktavatsalam, G., A.P.K. Reddy and V.T. John. 1977. Chemical control of sheath blight of rice. *Pesticides*, **11**: 13-16.

Bhaktavatsalam, G., K. Satyanarayana, A.P.K. Reddy and V.T. John, 1978. Evaluation of sheath blight resistance in rice. *Int. Rice Res. Newsl.*, **3** : 10.

Boogert, P.H.J.F. van den and J.W. Deacon, 1994. Biotropic mycoparasitism by *Verticillium biguttatum* on *R. solani*. *European J. Pl. Path.*, **100** : 137-156.

Boyatee, C.D. and F.N. Lee. 1979. Reduction in rice yield caused by sheath blight. *Arkans. For. Res.*, 28 : 5.

Burt, E.A. 1926. The Thelephoraceae of North America XV. *Corticium Ann. Mo. Dot. Gdn.*, 23 : 173-354.

Campbell, R. 1989. *Biological control of microbial plant pathogens*. Cambridge Univ. Press, Cambridge, p. 217.

Chen, Y.S. 1958. Studies on metabolic products of *Hypochnus sasakii* Shirai. Isolation of *p*-hydrodyphenylacetic acid and its physiological activity. *Bull. Agric. Chem. Soc. Japan*, No. 22 : 136-142.

Chien, C.C. and C.L. Chu. 1973. Studies on the control of rice blast and sheath blight of rice with benlate. *J. Taiwan Agric. Res.*, **22** : 41-46.

Chien, C.C. amd S.C. Chung. 1963. Physiologic races of *Pellicularia sasakii* in Taiwan. *J. Taiwan Agric. Res.*, **12** : 1-6.

Chien, C.C., Y.C. Hung and T.N. Lin. 1969. Studies on the pathogenicity of sclerotia of rice sheath blight fungus, *Pellicularia sasakii*. *J. Taiwan Agric. Res.*, **18** : 1-5.

Chin, K.M. and W.C. Lim. 1975. The survival, spread and host range of *Rhizoctonia solani*, the causal organism of sheath blight disease in rice. *MARDI* (*Malays. Agric. Res. Dev. Inst.*) *Res. Bull.*, **3** : 19-24.

Chu, V. and Y.L. Chen. 1964. Control of sheath blight of rice by 'K.C.I. 911'. *Acta Phytophylac. Sin.*, **3** : 1-14.

Coley-Smith, J.R. and R.C. Cooke. 1971. Survival and germination of fungal sclerotia. *Rev. Phytopathol.*, **9** : 65-92.

Cook, R.J. 1989. Biological control and holistic plant-health care in agriculture. *Am. J. Alt. Agric.*, **3** : 51-62.

Crill, P., F.L. Nuque, B.A. Estrad and J.M. Bandong. 1981. The role of varietal resistance in disease management. *Evaluation of the gene rotation concept for rice blast control*. IRRI, Box 933, Manila, Philippines. pp. 103-121.

CRRI, Cuttack, India. 1980. *A. Rep*. 1979.

CRRI, Cuttack, India. 1981. *A. Rep*. 1980.

CRRI, Cuttack, India. 1982. *A. Rep*. 1981.

Das, L. and S. Balakrishnan. 1978. Anastomosis grouping of cowpea isolates of *Rhizoctonia solani*. *Indian Phytopath.*, **36** : 200.

Das, N.P. 1970. Resistance of some improved varieties of rice (*Oryza sativa*) to sheath blight caused by *Rhizoctonia solani* Kühn. *Ind. J. Agric. Sci.*, **40** : 566-568.

Das, S.C. and S. Panda. 1984. Chemical control of rice sheath blight disease. *Indian Phytopath.*, **37** : 79-84.

Das, S.R. 1975. Occurrence of sheath blight disease of rice in western Orissa. *Indian J. Mycol. Pl. Path.*, **5** : 54.

Dasgupta, M.K. 1992. Rice sheath blight : The challenge continuous. In : *Plant diseases of International Importance*. Vol. I. Singh, U.S., A.N. Mukopadhayay, J. Kumar and H.S. Chanbe (eds.). Printice Hall, New Jersey. p. 130-157.

Dath, A.P. 1982. Effect of soil amendment with some green manures on the survival of sclerotia of *Corticium sasakii*. *Indian Phytopath.*, **35** : 523-525.

Dath, A.P. 1984. Effect of some cultural factors on the aggressiveness of *Rhizoctonia solani* on rice. *Indian Phytopath.*, **37** : 469-472.

Dath, A.P. 1985a. Rice chaff and rachillae - potential inoculum sources for sheath blight infection. *Curr. Sci.*, **54** : 432-433.

Dath, A.P. 1985b. A better criterion in rating the reaction of rice cultivars against sheath blight. *Indian Phytopath.*, **38** : 678-682.

Dath, A.P. 1987. A modified multipurpose detached leaf technique for rice sheath blight investigations. *Curr. Sci.*, **56** : 269-270.

Dath, A.P. 1988. Infectivity and variability of different inocula used for *Rhizoctonia solani* inoculation in upland rice. *Indian Phytopath.*, **41** : 136-138.

Dath, A.P. 1990. *Sheath blight disease of rice and its management.* Associate publishing company, New Delhi. p. 152.

Dath, A.P. and A. Swain. 1979. *In vitro* effect of certain nematicides and weedicides on *Corticium sasakii. Indian J. Mycol. Pl. Path.*, **9** : 95-96.

Dev, V.P.S. and C.A. Mary. 1983. Varietal resistance to leaf blast and sheath blight at Pattambi, India. *Int. Rice Res. Newsl.*, **8** : 5-6.

DRR-Directorate of Rice Research, Hyderabad, India. 1986. *A. Rep.* 1985.

Dodman, R.L. and T.T. Flentje. 1970. The mechanisms and physiology of plant penetration by *Rhizoctonia solani.* In : *Rhizoctonia solani, Biology* and *Pathology* (ed.) J.R. Parmeter Jr. Univ. Calif. Press, Berkley, Los Angels and London, pp. 149-160.

Dorairajan, C.P. 1993. Plant associated bacterial colonization in the rice system in relation to *Rhizoctonia* sheath blight control. Ph. D. Thesis. Andhra Pradesh Agric. Univ., Hyderabad, p. 134.

Eguchi, J. and seven others. 1973. Control of rice sheath blight by new polyoxin formulations and their practical properties. *Pro. Assoc. Pl. Prot. Hokuriku,* **21** : 77-81.

Endo, S. 1927. Ueber die Sclerotienkrankheit der Reispflanzen in den Phillippinen. *Pl. Prot. Tokyo.*, **14** : 283-288.

Endo, S. 1930. On the influence of the temperature upon the development of *Hypochnus. Ann. Phytopathol. Soc. Japan*, **2** : 280-283.

Endo, S. 1931. Studies on *Sclerotium* diseases of the rice plant. V. Ability of overwintering of certain important fungi causing *Sclerotium* disease of the rice plant and their resistance to dry conditions. *Forschin. Geb. Pflkrankh.*, **1** : 149-167.

Endo, S. 1933. Influence of salt on the pathogenicity of *Hypochnus sasakii* (Shirai). *Trans. Tottori Soc. Agric. Sci.*, **4** : 362-367.

Endo, S. 1935. On the influence of hydrogen-ion concentration on the mycelial growth of the causal fungi of sclerotial diseases of the rice plant. *Bull. Miyazaki Coll. Agric. For.*, **8** : 1-11.

Endo, S., M. Shinohara and S. Hara. 1965. Effects of PCP-Na herbicide on sheath blight of rice plants. I. The inhibitive mechanism of the PCP-Na herbicide to the occurrence of the diseasse. II. Residual effectiveness of PCP-Na herbicide on *Pellicularia sasakii* (Shirai) S. Ito in irrigation. *Bull. Coll. Agric. Vet. Med. Nihon Univ.*, No. **20** : 1-9, 11-17.

Endo, S., M. Shinohara and K. Yamane. 1974c. Studies on the causal fungus of summer blight of cirmson clover. I. Comparison between the soil-inoculated causal fungus of summer blight of crimson clover and that of sheath blight of rice plant (*Corticium sasakii* (Shirai) Matsumoto) in pathogenicity to some forage grasses and legumes and rice plant. *Bull. Coll. Agric. Vet. Med. Nihon Univ,*. No. **31** : 124-133.

Endo, S., M. Shinohara, M. Kohata and T. Kanehira. 1974a. Studies on the causal fungus of summer blight of crimson clover. II. Comparison of the causal fungus of summer blight of crimson clover with that of sheath blight of rice plant (*Corticium sasakii* (Shirai) Matsumoto) and *Rhizoctonia solani* Kühn in pathogenicity to rice plant and crimson clover, and types of disease spots formed on the inoculated plants. *Bull. Coll. Agric. Vet. Med. Nihon Univ.*, No. **31** : 150-165.

Endo, S., M. Shinohara, Y. Kobayashi and M. Hiramatsu. 1973. Effect of the antagonistic action of *Neurospora crassa* Shear & Dodge on the occurrence of sheath blight of the rice plant caused by *Pellicularia sasakii* (Shirai) S. Ito. *Bull. Coll. Agric. Vet. Med. Nihon Univ.*, No. **30** : 96-105.

Endo, S., M. Shinohara, M. Maki, K. Suzuki and Y. Kobayashi. 1974b. Studies on the causal fungus of summer blight of crimson clover. III. Comparison of the causal fungus of summer blight of crimson clover with that of sheath blight of rice plant (*Corticium sasakii* (Shirai) Matsumoto) and *Rhizoctonia solani* Kühn in some cultural characters and sclerotial morphology. *Bull. Soll. Agric. Vet. Med. Nihon Univ.*, No. **31** : 134-139.

Endo, S. and T. Kanehira, M. Shinohara, T. Sasaki, H. Omori and T. Takemori. 1976. Comparative studies of the causal fungi of sheath blight of rice plant (*Corticium sasakii* (Shirai) Matsumoto) and black scurf of potato plant (*Rhizoctonia solani* Kühn) II. Some cultural characters on artificial media. *Bull. Coll. Agric. Vet. Med. Nihon Univ.*, No. **33** : 168-185.

Flinker, A., Y. Koltin, I. Barash, B. Sneh and D. Pozniak. 1985. Isolation of virus from virulent strain of *Rhizoctonia solani. J. Gen. Virology,* **66** : 1221-1232.

Fukatsu, R., T. Kakijiki and S. Hirayama. 1960. Studies on secondary infection by basidiospores of sheath blight fungus. *Bull. Koochi. Agric. Exp. Stn.*, No. **2** : 26-32.

Furuta, H. 1973. Validamycin (trade name : Validacin), a new antibiotic for sheath blight of rice. *Japan Pest. Inf.*, **15** : 28-30.

Gadd, C.H., and L.S. Bertus. 1928. *Corticium vagum* B and C - Cause of disease of *Vigna oligosperma* and other plants of Ceylon. *A Rev. Bot. Gdns. Peradeniya*, **11** : 27-49.

Gangopadhyay, S. 1983. *Current concepts on fungal diseases of rice*. Today and Tomorrow's Printers and Publishers, New Delhi, p. 349.

Gangopadhyay, S. and N.K. Chakrabarti. 1982. Sheath blight of rice. *Rev. Pl. Path.*, 61 : 451-460.

Gangopadhyay, S. and K.M. Das, 1982. Occurrence of vesicular-arbuscular mycorrhizae in rice in India. *Ind. Phytopath.*, **35** : 83-85.

Gangopadhyay, S. and K.M. Das, 1988. Control of soil-borne diseases of rice through vesicular-arbuscular mycorrhizae. In Proc. Natl. workshop on mycorrhizae. Jawaharlal Nehru Univ., New Delhi, pp. 560-580.

Gerdemann, J.W. 1968. Vesicular-arbuscular mycorrhiza and plant growth. *Ann. Rev. Phytopathol.*, **6** : 397-418.

Gilligan, C.A. 1983. Modeling of soilborne pathogens. *Ann. Rev. Phytopathol*, **21**: 45-64.

Gnanamanickam, S.S., B.L. Candole and T.W. Mew. 1992. Influence of soil factors and cultural practice on biocontrol of sheath blight of rice with antagonistic bacteria. *Plant and Soil*, **144** : 67-75.

Gokulapalan, C. and M.C. Nair. 1983. Collateral hosts of *Rhizoctonia solani* Kühn causing sheath blight of rice *Int. Rice Res. Newsl.*, **8** : 10.

Gokulapalan, C. and M.C. Nair. 1984. Antagonism of a few fungi and bacteria against *Rhizoctonia solani* Kühn. *Indian. J. Microbial*, **24** : 57-58.

Gokulapalan, C. and M.C. Nair. 1986. Mycoparasites of *Rhizoctonia solani* and control of sheath blight or rice. 14th Annual meeting of the mycological society of India and seminar on applied mycology. Thanjavur, India. p. 18. (Abstr.).

Greenland, D.J. and S.K. De Datta 1985. Constraints to rice production and wetland soil characteristics. In : Wet soils: Characterization and utilization. IRRI, Box 933, Manila, Philippines, pp. 23-36.

Guo, C.J., Z.Y. Chen and F.M. Wang. 1985. Pathogenic variability of sheath blight of rice caused by *Thanatephorus cucumeris* (Frank) Donk and identification techniques of varietal resistance. *Sc. Agric. Sin.*, **5** : 50-57.

Hadar, Y., Chet. I., and Henis, Y. 1979. Biological control of *Rhizoctonia solani* damping-off with wheat bran culture of *Trichoderma harzianum. Phytopathology*, **69**: 64-68.

Haque, M.A. 1975. Variation in cultural characteristics, physiology, and pathogenicity of field isolates of *Thanateporus cucumeris* (Frank) Donk, the causal organism for rice sheath blight disease. M.S. Thesis, Univ. of the Philippines, Los Baños, Laguna, Philippines, p. 79.

Haque, M.A., S.H. Ou and H.E. Kauffman. 1977. Response of rice sheath blight fungus to higher concentration of carbon dioxide in growth tube. *Bangladesh. J. Agric.*, **1** : 135-137.

Harman, G.E. 1988. New strategies of control of soil borne plant pathogens by non-conventional means. In : 5th *Intl. Cong. of Pl. Path. Kyoto, Japan*, p. 165.

Hasama, W. 1989. Occurrence of sheath blight disease of rice and its control of Oita prefecture. *Noyaku Kenkyu*, **35** : 21-26.

Hashiba, T. 1982a. Sclerotial morphogenesis in the rice sheath blight fungus (*Rhizoctonia solani*). *Bull. Hokuriku Natl. Agric. Expt. Stn.*, **24** : 29-83.

Hashiba, T. 1982b. Formation and fusion of protoplasts from *Rhizoctonia solani*. *Jap. J. Med. Mycol.* **24** : 143-150.

Hashiba, T. 1984a. Estimation method of severity and yield loss by rice sheath blight disease. *Bull. Hokuriku Natl. Agric. Expt. Stn.*, **26** : 155-164.

Hashiba, T. 1984b. Forecasting model and estimation of yield loss by rice sheath blight disease. *J. Agric. Res. Quat.*, **18** : 92-98.

Hashiba, T. 1985a. Ecological studies of rice sheath blight disease. *Ann. Phytopathol. Soc. Japan*, **51** : 252-254.

Hashiba, T. 1985b. New forecasting model of yield loss by rice sheath blight disease. *Plant Prot.*, **39** : 332-338.

Hashiba, T. and T. Ijiri. 1989. Estimation of yield loss and computerized forecasting system (BLIGHTAS) for rice sheath blight disease. *Trop. Agric. Res. Centre Ser.*, No. **22** : 163-172.

Hashiba, T. and S. Mogi. 1973. The number of germination ability of sclerotia of the rice sheath blight fungus. *Phytopathology,* **65** : 160-162.

Hashiba, T. and S. Mogi. 1975. Developmental changes in sclerotia of the rice sheath blight fungus. *Phytopathology,* **65** : 160-162.

Hashiba, T. and R.C. Staples. 1976. Nucleic acid metabolism in developing sclerotia of the rice sheath blight fungus (*Rhizoctonia solani* Kühn). *J. Gen. Microbiol.,* **96** : 239-245.

Hashiba, T. and M. Yamada. 1981. Viability of sclerotia formed on rice plants and on culture media by *Rhizoctonia solani. Ann. Phytopathol. Soc. Japan,* **47** : 464-471.

Hashiba, T. and Yamada, M. 1982. Formation and purification of protoplasts from *Rhizoctonia solani. Phytopathology,* **72** : 849-853.

Hashiba, T. and T. Yamaguchi. 1971. Effect of temperature and humidity on the germination of sclerotium of *Pellicularia sasakii* (Shirai) S. Ito *Proc. Assoc. Pl. Protect. Hokuriku,* pp. 6-10.

Hashiba, T., K. Koike and M. Yamada. 1983. A method for estimation of the yield loss of rice sheath blight disease based on height of the lesions and percentage of affected hills in rice. *Ann. Phytopathol. Soc. Japan,* **49** : 143-147.

Hashiba, T., S. Mogi and S. Yagi. 1974. The relation between the mycelial growth of rice sheath blight fungus isolates and the air temperature of the collecting region. *Proc. Assoc. Pl. Protect. Hokuriku,* **22** : 8-14.

Hashiba, T., H. Uchiyamada and K. Kimura. 1981. A method to estimate the disease incidence based on the height of infected part in rice sheath blight disease. *Ann. Phytopathol. Soc. Japan,* **47** : 194-198.

Hashiba, T., T. Yamaguchi and S. Mogi. 1972. Biological and ecological studies on the sclerotium of *Pellicularia sasakii* (Shirai) S. Ito. I. Floating on the water surface of sclerotium. *Ann. Phytopathol. Soc. Japan,* **38** : 414-425.

Hashiba, T., T. Yamaguchi and S. Mogi. 1977. Quantitative changes in nitrogen and starch contents in rice sheaths during vertical disease development of sheath blight caused by *Rhizoctonia solani* Kühn. *Ann. Phytopathol. Soc. Japan,* **43** : 1-8.

Hashiba, T., K. Koike, I. Yuno and M. Yamada. 1983. Mode of vertical development of sheath blight lesions on rice plants caused by *Rhizoctonia solani* Kühn. *Ann. Phytopathol. Soc. Japan,* **48** : 449-509.

Hashiba, T., Y. Homma, M. Hyakumachi and I. Matsuda. 1984. Isolation of a DNA plasmid in the fungus *Rhizoctonia solani. J. Gen. Microbiol.,* **130** : 2067-2070.

Hashioka, Y. 1951a. Varietal resistance of rice sheath blight and the sclerotial diseases (Studies on pathological breeding of rice, IV). *Japan J. Breed.,* **1** : 21-26.

Hashioka, Y. 1951b. Inheritance of resistance to sheath blight in rice varieties. *Ann. Phytopathol. Soc. Japan,* **15**: 98-99.

Hashioka, Y. 1956. Prevalence and fungicidal control of the rice sheath blight. *Agric. Hort. Tokyo,* **31** : 953-957.

Hashioka, Y. and T. Fukita. 1969. Ultrastructural observations on mycoparasitism of *Trichoderma, Gliocladium* and *Acremonium* to pathogenic fungi. *Rep. Tottori. Mycol. Inst.,* **7** : 8-18.

Hashioka, Y. and M. Makino. 1960. Phytopharmacological studies on the rice diseases. VII. Tolerance of sclerotia of rice sheath spot fungus, *Corticium sasakii,* to organo mercurials and organo arsenicals. *Res. Bull. Fac. Agric. Gifu Univ.,* No. **12** : 36-44.

Hashioka, Y. and M. Makino. 1961. Examination of methods of chemical application in irrigation water in paddy field against sheath blight disease of rice plant. *Proc. Kansai Pl. Prot. Soc.,* **3** : 25-29.

Hashioka, Y. and M. Makino. 1969. *Rhizoctonia* group causing the rice sheath spots in the temperate and tropical regions, with special reference to *Pellicularia sasakii* and *Rhizoctonia oryzae. Res. Bull. Fac. Agric. Gifu Univ.,* No. **28** : 51-63.

Hashioka, Y. and K. Okuda. 1971. Scanning and transmission electron microscopy on the initial infection of rice sheath by *Pellicularia sasakii* and *Rhizoctonia oryzae. Res. Bull. Fac. Agric. Gifu Univ.,* No. **31** : 99-111.

Hemmi, T. and S. Endo. 1928. On the staining method for testing the viability of sclerotia of fungi. *Mem. Coll. Agric. Kyoto Univ.,* **7** : 39-49.

Hemmi, T. and S. Endo. 1931. Studies on *Sclerotium* diseases of the rice plant. III. Some experiments on the sclerotial formation and the pathogenicity of certain fungi causing sclerotium diseases of the rice plant. *Forsch. Geb. Pflkrankh.,* **1** : 111-125.

Hemmi, T. and S. Endo. 1933. Studies on *Sclerotium* diseases of the rice plant. VI. On the relation of temperature and period of continuous wetting to the infection of rice plant by *Hypochnus sasakii* Shirai. *Forsch. Geb. Pflkrankh.*, **2** : 202-218.

Hemmi, T. and K. Yokogi. 1927. Studies on *Sclerotium* diseases of the rice plant. I. *Agric. Horti. Tokyo*, **2** : 955-1094.

Hirooka, T., Y. Miyagi and F. Araki. 1988. Effect of flutolanil on the infection process of *Rhizoctonia solani*. In Symposium, IV. Development of new fungicides active against basidiomycetes. 5th *Int. Cong. Pl. Path., Kyoto, Japan*, pp. 22 (Abstr.).

Hirooka, T., H. Kasahara, Y. Miyagi and H. Kunoh. 1990. Effect of the systemic fungicide flutolanil on morphology of *Rhizoctonia solani* following inhibition of succinate oxidation. *J. Pesticide Sci.* **15** : 47-53.

Hirooka, T. Y. Miyagi, F. Araki and H, Kunoh. 1989. Biological mode of action of flutolanil in its systemic control of rice sheath blight disease. *Phytopathology*, **79** : 1091-1094.

Hori, M. 1969. On forecasting the damage due to sheath blight of rice plants and the critical point for judging the necessity of chemical control of the disease. *Rev. Pl. Prot. Res.*, **2** : 70-73.

Hori, M. and M. Anraku. 1971. Studies on the forecasting techniques of sheath blight of rice plant. Appointed Experiment (Plant diseases and insect pests) *Yamaguchi Agric. Res. Stn. Spl. Bull.*, No. **24** : 1-85.

Hsieh, Y.T., Y.L. Wu and K.N. Shian. 1965. Screening for sheath blight resistance in rice varieties. *Kaohsiung Dist. Agric. Imp. Stn. Taiwan*, p. 92.

Hsu, S.C. and T.C. Dough. 1964. Effect of chemical elements on the growth of crop pathogens. II. Effect of Hg-, Ni-, Sn-, Pb-, Fe- and Al- salts on rice blast and sheath spot fungi. *J. Taiwan Agric. Res.*, **3** : 38-43.

Iguchi, K. 1988. Effective application time of chemicals for the control of rice sheath blight in Chiba prefecture. *Proc. Kanto-Tosan Pl. Prot. Soc. Japan*, **35** : 14-15.

Ijiri, T. and T. Hashiba. 1986. Computerized forecasting system for rice sheath blight disease (BLIGHTAS) *J. Pl. Prot.*, **40** : 42-45.

Ikata, S. and T. Hitomi. 1930. On the mode of primary infection through sclerotia and field observations on the basidiosphore formation in the earlier stage of Monko disease (*Hypochnus Sasakii* Shirai) of rice plants. *J. Pl. Prot. Tokyo*, **17** : 17-28.

Imolehin, E.D. 1989. Rice diseases in the Southern Guinea Savannah Zone of Niger State, Nigeria. *Int. Rice Res. Newsl.*, **14** : 37-38.

Inagaki, K. 1983. Sheath blight and the other rice sclerotial diseases caused by *Rhizoctonia oryzae* and 3 *Sclerotium* species with respect to their concurrent occurrences in nature and on rice plants. *Ann. Phytopathol. Soc. Japan*, **49** : 736-738.

Inagaki, K. and M. Makino. 1974. Karyological characters of the fungi causing sclerotiosis. *Ann. Phytopathol. Soc. Japan*, **40** : 368-371.

Inagaki, K. and M. Nakamoto. 1982. Occurrence trends of sclerotiosis caused by *Rhizoctonia oryzae* Ryker et Gooch and 3 spp. of sclerotium on rice in the south western area of the main island and Nansei-shoto island of Japan. *Sci. Rep. Fac. Agric. Meijo Univ.*, **18** : 20-24.

Inoue, Y. and K. Uchino. 1963. Studies on sheath blight of rice plant caused by *Pellicularia sasakii* (Shirai) S. Ito. I. Ecology of damage and chemical control. *Ministry of Agric. Forestry and Fisheries Res. Coun. Jap. App. Exp. No. 4 at Yamaguchi Agric. Exp. Stn.*, p. 136.

IRRI. 1971. *A. Rep*. 1970.

IRRI. 1972. *A. Rep*. 1971.

IRRI. 1973. *A. Rep*. 1972.

IRRI. 1978. *A. Rep*. 1977.

IRRI. 1985. *A. Rep*. 1984.

IRRI. 1987. *A. Rep*. 1986.

IRRI. 1988. SES for rice. 3rd ed. p. 54.

IRRI. 1989. *A. Rep*. 1988.

Isawa, T., E. Higashide, H. Yamamoto and M. Shibata. 1971. Studies on validamycins, new antibiotics. II. Production and biological properties of validamycin A and B. *J. Antibiotics*, **24** : 107-133.

Jeyarajan, R. and S. Nakkeeran. 1995. Exploitation of biocontrol potential of *Trichoderma* for field use. In : *Recent developments in biocontrol of plant pathogens* (K. Manibhushanrao and A. Mahadevan eds.), Today and Tomorrow's Printers and Publishers, New Delhi (in press).

Johnson, K.B. 1994. Dose-response relationships and inundative biological control. *Phytopathology*, **84** : 780-784.

Kalaiselvi, K., S. Sreenivasaprasad and K. Manibhushanrao. 1986. Acquired resistance in rice leaves to *Rhizoctonia solani*. *Int. Rice. Res. Newsl.*, **11** : 16.

Kamjaipai, W. and P. Giatgong. 1971. Number of nuclei of the fungus causing sheath blight of rice, *Corticium* spp., for taxonomic position. *Thailand J. Agric. Sc.*, **4** : 71-78.

Kang, I.M., E.K. Lee, S.C. Lee and Y.K. Lim. 1965. Studies on the resistance of rice variety to sheath blight (*Corticium sasakii*) in the field. *Res. Rep. Korea Rural Develop.*, **8** : 235-241.

Kannaiyan, S. and N.N. Prasad. 1976. Efficacy of certain fungicides on the control of sheath blight. *Int. Rice Res. Newsl.*, **1** : 19.

Kannaiyan, S. and N.N. Prasad. 1978. Seed-borne nature of sheath blight pathogen *Rhizoctonia solani* in rice. *Int. Rice Res. Newsl.*, **3** : 10.

Kannaiyan, S. and N.N. Prasad. 1978a. Estimation of yield losses due to sheath blight disease of rice. *Aduthurai Rep.*, **2** : 137-138.

Kannaiyan, S. and N.N. Prasad. 1978b. Control of sheath blight disease of rice with Aureofungin. *Hindustan Antibiotics Bull.*, **24** : 31-37.

Kannaiyan, S. and N.N. Prasad. 1979a. Varietal resistance of rice sheath blight. *Int. Rice. Newsl.*, **4** : 4.

Kannaiyan, S. and N.N. Prasad. 1979b. Influence of nitrogen fertilization on development of sheath blight. *Int. Rice Ress. Newsl.*, **4** : 14.

Kannaiyan, S. and N.N. Prasad. 1979c. Sheath blight incidence in weed hosts. *Int. Rice Res. Newsl.*, **4** : 17.

Kannaiyan, S. and N.N. Prasad. 1980. Effect of certain phenolic compounds on the growth and sclerotial production of *Rhizoctonia solani*. *Phytopath. Z.*, **98** : 178-181.

Kannaiyan, S. and N.N. Prasad. 1981a. Changes in the rhizosphere microflora of rice on infection with *Rhizoctonia solani*. *Oryza*, **18** : 246-247.

Kannaiyan, S. and N.N. Prasad. 1981b. Effect of organic amendments on seedling infection of rice caused by *Rhizoctonia solani*. *Pl.* and *Soil*, **62** : 131-133.

Kataria, H.R. and U. Gisi. 1989. Recovering from soil and sensitivity to fungicides of *Rhizoctonia cerealis* and *R. solani*. *Mycol. Res.*, **92** : 458-462.

Kawakubo, Y. 1984. Control effect of moncern (Pencycuron) for sheath blight disease in Fukui prefecture. *Noyaku Kenkyu*, **31** : 12-16.

Kernkamp, M.F. 1952. Investigations on physiological specialization and parasitism of *Rhizoctonia solani*. *Univ. Minnesota Agric. Expt. Stn. Tech. Bull.*, **200** : 36.

Khan, A.H., A. Islam, R. Islam, S. Begum and S.M.I. Huq. 1988. Mycorrhizal status of some Bangladesh soils and the effect of indigenous VA mycorrhizal fungi on the growth of rice plants. *Bangladesh J. Bot.*, **17** : 49-56.

Khatua, D.C. 1982. Weed host of *Rhizoctonia solani* Kühn a rice sheath blight pathogen. *Int. Rice Res. Newsl.*, **7**: 13.

Kim, C.H. and C.K. Kim. 1987. Density and viability of sclerotia of rice sheath blight pathogen overwintering in field. *Korean J. Pl. Prot.*, **26** : 99-106.

Kim, C.H. and C.K. Kim. 1988. Evaluation for duration of fungicidal effect in rice sheath blight. *Res. Rept. RDA.*, **30** : 1-4.

Kim, C.H. and H.S. Min. 1983. Ecological studies on rice sheath blight caused by *Rhizoctonia solani*. II. Forecasting and control of rice sheath blight. *Korean J. Pl. Prot.*, **22** : 21-25.

Kim, C.K., D.S. Ra, H.S. Min and Y.H. Lee. 1985. Ecological studies on rice sheath blight caused by *Rhizoctonia solani*. III. Cultural method and disease development. IV. Inoculation time of sclerotium and disease development. *Korean J. Pl. Pathol.*, **2** : 7-10; **3** : 169-172.

Kim, C.K., C.S. Kang, S.J. Rim, E.J. Lee and K.H. Lee. 1981. Ecological studies on rice sheath blight caused by *Rhizoctonia solani*. I. The difference of disease development between Jinneung and Yushin. *Korean J. Pl. Prot.*, **20** : 71-75.

Kim, K.H., J.B. Park and K.J. Yang. 1990. Studies on varietal resistance to sheath blight disease in rice. V. Relationship between infection structure formation and disease severity. *Korean J. Breed.*, **22** : 13-19.

Kim, W.G. and C.K. Kim. 1988. Density of overwintered sclerotia in paddy fields in Korea, viability of the sclerotia and pathogenicity of the sclerotial fungi. *Korean J. Pl. Pathol.*, **4** : 207-217.

Kim, W.G. and R. Yoshino. 1988. Characteristics in hyphal anastomosis of *Rhizoctonia solani*. Ag-1 (A) in relation to locality and cultural ages of the isolates. *Ann. Phytopathol. Soc. Japan,* **54** : 141-150.

Kim, W.G., C.K. Kim and D.S. Ra. 1989. Development of four sclerotial disease of rice and effect of Neoasozin Lq in the paddy field. *Res. Rep. RDA, Sewon, Korea,* **31** : 1-6.

Kim, W.G., C.K. Kim and S.H. Yu. 1988. Anastomosis grouping and cultural characteristics of isolates of *Rhizoctonia solani* Kühn from sclerotia overwintered in paddy fields. *Korean J. Pl. Pathol.*, **4** : 136-142.

Ko, W.H. and F.K. Hora. 1971. Selective medium for the quantitative determination of *Rhizoctonia solani* in soil. *Phytopathology,* **61** : 701-707.

Kohil, C.K. 1966. Pathogenicity and host range studies of paddy sheath blight pathogen (*Rhizoctonia solani* Kühn). *J. Res.* PAU, *Ludhiana,* **3** : 37-40.

Kohmoto, K. and S. Nishimura. 1974. Pathochemical studies on *Rhizoctonia* diseases. II. Substrate specificity in specific meta-hydroxylation of aromatic acids by *Rhizoctonia solani* Kühn. *Ann. Phytopathol. Soc. Japan,* **40** : 79-85.

Kohmoto, K., R. Fukui, M. Mizuno and S. Nishimura. 1973. Pathochemical studies on *Rhizoctonia* diseases III. Characteristic production of *Rhizoctonia* toxins and their *vivo*-toxicity. *J.Fac. Agric. Tottari Univ.*, **8** : 21-31.

Konishi, K. and T. Kuragano. 1989. Fungicidal activity of aromatic ketone pyrimidinyl-hydrazones. *J. Pestic. Sci.*, **14** : 211-221.

Kormanik, P.P., W.C. Bryan and R.C. Schultz. 1980. Procedures and equipment for staining large numbers of plant root samples for endomycorrhizal assay. *Can. J. Microbiol.*, **26** : 536-538.

Kozaka, T. 1961. Ecological studies on sheath blight of rice plant caused by *Pellicularia sasakii* (Shirai) S. Ito. and its chemical control. *Chuguku Agric. Res.*, **20** : 1-132.

Kozaka, T. 1965. Ecology of *Pellicularia* sheath blight of rice plant and its chemical control. *Ann. Phytopathol. Soc. Japan,* **31** : 179-185.

Kozaka, T. 1970. *Pellicularia* sheath blight of rice plant and its control. *Japan Agric. Res.*, **5** : 12-16.

Kozaka, T. 1975. Sheath blight of rice plants and its control. *Rev. Pl. Prot. Res.*, **8** : 69-80.

Kuan, T.L. and M.S. Kuo. 1974. Production of phosphatidase by *Thanatephorus cucumeris* (Frank) Donk and its relation to pathogenesis of rice sheath blight disease. *Pl. Prot. Bull. Taiwan,* **16** : 1-11.

Kuck, K.H., I. Ueyama, S. Kurogochi and Y. Yamada. 1988. Biological and biochemical aspects of (R) Moncern (Pencycuron), a *Rhizoctonia* specific fungicide. In : Symposium IV. Development of new Fungicides active against basidiomycetes. 5th *Int. Cong. Pl. Path., Kyoto, Japan,* p. 22 (Abstr.).

Kumaresen, S. and K. Manibhushanrao. 1991. Studies on the biological control of rice sheath blight disease. *Indian J. Pl. Pathol.*, **9** : 64-70.

Kundu, P.K. and B. Nandi. 1985. Control of *Rhizoctonia* disease of cauliflower by competetive inhibition of the pathogen using organic amendments in soil. *Pl. and Soil,* **83** : 37-362.

Kuninaga, S. and R. Yokasawa. 1984a. DNA base sequence homology in *Rhizoctonia solani* Kühn. IV. Genetic relatedness within AG-4. *Ann. Phytopathol. Soc. Japan,* **50** : 322-330.

Kuninaga, S. and R. Yokasawa. 1984b. DNA base sequence homology in *Rhizoctonia solani* Kühn. V. Genetic relatedness within AG-6. *Ann. Phytopath Soc. Japan,* **50** : 346-352.

Kuninaga, S., R. Yokosawa and A. Ogoshi. 1979. Some properties of anastomosis groups 6 and Bl in *Rhizoctonia solani* Kühn. *Ann. Phytopathol. Soc. Japan,* **45** : 207-214.

Kurodani, K., K. Yokogi and M. Yamamoto. 1959. On the effect of 2, 4-dichlorophenoxyacetic acid on the mycelial growth of *Hypochnus sasakii* Shirai. *Forschn. Geb. Pflkrankh. Kyoto,* **6** : 132-135.

Lakshmanan, P. and M.C. Nair. 1980. *In vitro* toxicity of granular insecticides against *Rhizoctonia solani. Madras Agric. J.*, **61** : 59-60.

Lana, S.K. and D.C. Khatua. 1984. *Axonopus campressus,* a grass host for *Rhizoctonia solani. Int. Rice Res. Newsl.*, **9** : 21.

Lee, F.N. 1980. Number, viability and buoyancy of *Rhizoctonia solani* sclerotia in Arkansas rice fields. *Pl. Dis.*, **64** : 298-300.

Lee, F.N. and M.L. Courntney. 1982. Burning rice straw compliments other sheath blight control tactics. *Arkansas Farm Res.*, **31** : 7.

Lee, F.N. and M.C. Rush. 1983. Rice sheath blight : A major rice disease *Pl. Dis.*, **67** : 829-832.

Lin, S.D. 1976. Association of pectin transeleminase in the pathogenesis of sheath blight disease of rice plant. *Natl. Sci. Coun. Mon.*, **4** : 32-41.

Loo, E.P., C.Q. Choun and D.C. Lee. 1963. Studies on *Rhizoctonia* blight of rice. *Acta Phytophylac. Sin.*, **2** : 431-440.

Malaguti, G. 1951. Machan de la hoja envariradora del Arroz causada por *Rhizoctonia solani* (Sheath spot disease of rice caused by *R. solani*). *Agronomia Trop. Maracay.*, **1** : 71-75.

Manian, S. 1982. Studies on the sheath blight disease of rice. Ph.D. Thesis, Univ. of Madras, Madras. p. 121.

Manian, S. and K. Manibhushanrao. 1980. The efficacy of the scale by the normal distribution method in screening rice cultivars for sheath blight disease. *Riso.*, **29** : 313-316.

Manian, S. and K. Manibhushanrao. 1982a. Histopathological studies in rice sheath blight incited by *Rhizoctonia solani. Z. Pflkrankh. Pflschutz.*, **89** : 523-531.

Manian, S. and K. Manibhushanrao. 1982b. Screening rice germplasm for tolerance to sheath blight incited by *Rhizoctonia solani* Kühn. *Phytopath. Medit.*, **21** : 37-38.

Manian, S. and K. Manibhushanrao. 1990. Influence of some factors on the survival of *Rhizoctonia solani* in soil. *Trop. Agric.*, **67** : 207-208.

Manibhushanrao, K. 1989. Sheath blight disease threatening rice cultivation in South East Asia. *J. Indian Bot. Soc.*, **68** : 10-18.

Manibhushanrao, K. 1990. Contribution to blast and sheath blight diseases of rice in India. D.Sc. thesis, Univ. of Madras, Madras, p. 126.

Manibhushanrao, K.1994. *Rice blast disease*, Daya Publishing House, Delhi, p. 180.

Manibhushanrao, K. and T. Mani. 1993. Glucosamine assay as a technique to detect fungal mass in rice-sheath blight system. *Z. Pflkrankh. Pflschutz.*, **100** : 285-290.

Manibhushanrao, K. and S. Manian. 1995. Changes in the phenol metabolism of rice cultivars with reference to nitrogen fertilization and sheath blight disease. *Acta phytopath. et Entamol. Hung.*, (In press).

Manibhushanrao, K. and P. Ramalingam. 1993. Studies on the phytotoxic metabolites of the rice sheath blight pathogen *Rhizoctonia solani. Indian J. Microbial Ecology,* **3** : 10-21.

Manibhushanrao, K. and S. Sreenivasaprasad. 1985. Inhibition of polygalacturonase activity of rice pathogens by the oxidized products of chlorogenic acid. *Curr. Sci.*, **54** : 146-148.

Manibhushanrao, K. and M. Zuber. 1978. Disease resistance in cereals. *Acta Phytopath. Hung.*, **13** : 313-335.

Manibhushanrao, K., U.I. Baby and Y. Joe. 1989a. Effect of organic amendments on the saprophytic survival of rice sheath blight pathogen and the soil microflora. *Oryza,* **26** : 71-89.

Manibhushanrao, K., D.M. Chelladorai and S. Sreenivasaprasad. 1995. Effect of phenylacetic acid on the phenolics and enzymes of phenol metabolism in rice. *Acta Phytopath. et Entamol. Hung.* (In press).

Manibhushanrao, K., Y. Joe and P. Madathiammal. 1990. Elicitation of resistance in rice to sheath blight disease. *Int. J. Trop. Pl. Dis.*, **8** : 193-197.

Manibhushanrao, K., S. Manian and M. Zuber. 1979. Sheath blight disease of rice in South East Asia. *Z. Pflkrankh. Pflschutz.*, **86** : 499-509.

Manibhushanrao, K., S. Manian and M. Zuber. 1991. The phytotoxic metabolites of *Rhizoctonia solani. J. Sci. and Ind. Res.*, **40** : 602-607.

Manibhushanrao, K., S. Sreenivasaprasad and V. Parvathi. 1986. Changes in phenolic constituents of rice in response to *Rhizoctonia solani* infection. *Indian J. Bot.*, **9** : 104-111.

Manibhushanrao, K., S. Sreenivasaprasad, U.I. Baby and Y. Joe. 1989b. Susceptibility of rice sheath blight pathogen to mycoparasites. *Curr. Sci.*, **58** : 515-518.

Manibhushanrao, K., S. Sreenivasaprasad, R.S. Chitralekha and K. Kalaiselvi. 1988. Cross protection in rice to sheath blight. *J. Indian Bot. Soc.*, **67** : 97-100.

Manibhushanrao, K., M. Zuber and N. Matsuyama. 1988. Phenol metabolism and plant disease resistance. *Acta Phytopath. et Entamol. Hung.*, **23** : 103-114.

Manibhushanrao, K., M. Zuber and S. Sreenivasaprasad. 1987. Mycelial amino acid composition of differentially virulent isolates of *Rhizoctonia solani*. *Curr. Sci.*, **56** : 619-621.

Marchetti, M.A. 1983. Potential impact of sheath blight on yield and milling quality of short-statured rice lines in the southern United States. *Pl. Dis.*, **67** : 162-165.

Marshall, D.S. and M.C. Rush. 1980a. Infection cushion formation on rice sheaths by *Rhizoctonia solani*. *Phytopathology*, **70** : 947-950.

Marshall, D.S. and M.C. Rush. 1980b. Relation between infection by *Rhizoctonia solani* and *R. oryzae* and disease severity in rice. *Phytopathology*, **70** : 941-946.

Martini, G., G. Gramaldi and I. Guardiola. 1978. Extrachromosal DNA in phytopathogenic fungi. In : *Genetic Engineering* (H.W. Boyer and S. Nicosia (eds.). Elsevier, North Holland, Amsterdam, pp. 177-200.

Mathai, G. and P.V. Nair. 1977. Field evaluation of fungicides against *Corticium sasakii* (Shirai) Matsumoto, causing sheath blight of rice. *Agric. Res. J. Kerala*, **14** : 184-186.

Matsumoto, T. 1934. Some remarks on the taxonomy of the fungus *Hypochnus sasakii* Shirai. *Trans. Sapporo Nat. Hist. Soc.*, **13** : 115-120.

Matsumoto, T., W. Yamamoto and S. Hirane. 1932. Physiology and parasitology of the fungi generally referred to as *Hypochnus sasakii* Shirai. I. Differentiation of the strains by means of hyphal fusion and culture in differential media. *J. Soc. Tropic. Agric.*, **4** : 370-388.

Matsumoto, T., W. Yamamoto and S. Hirane. 1933. Physiology and parasitism of the fungi generally referred to as *Hypochnus sasakii* Shirai. II. Temperature and humidity relations. *J. Soc. Tropic. Agric.*, **5** : 332-345.

Matsuura, K. 1986. Scanning electron microscopy of the infection process of *Rhizoctonia solani* in leaf sheaths of rice plants. *Phytopathology*, **76** : 811-814.

Matsuura, K., R. Shigemoto and T. Okuno. 1988. Biological properties and mode of action of validamycin A. In : Symposium IV. Development of new fungicides active against basidiomycetes. 5th *Int. Cong. Pl. Path., Kyoto, Japan*, p. 22 (Abstr.).

Matsuyama, N., Z. Moromizato, A. Ogoshi and S. Wakimoto. 1978. Grouping of *Rhizoctonia solani* Kühn with non-specific esterase zymogram. *Ann. Phytopathol. Soc. Japan*, **44** : 652-658.

Matsuyama, N., M. Yahara, S. Wakimoto and S. Kuboto. 1985. Inhibitory effect of 1, 3-Dithiol derivatives on the growth and sclerotium formation of *Rhizoctonia solani* Kühn (AG-1). *Ann. Phytopathol. Soc. Japan*, **51** : 595-598.

Mew, T.W. 1988. World situation of rice diseases. 5th *Int. Cong. Pl. Path. Kyoto, Japan*, p. 6 (Abstr.).

Mew, T.W. and A.M. Rosales. 1984. Relationship of soil microorganisms to rice sheath blight development in irrigated and dry land rice culture. In : *Soilborne diseases in Asia, food and fertilizer*. Tech. Centre, Taiwan Book Ser. No. **26**, pp. 147-158.

Mew, T.W. and A.M. Rosales. 1986. Bacterization of rice plants for control of sheath blight caused by *Rhizoctonia solani*. *Phytopathology*, **76** : 1260-1264.

Mew, T.W. and A.M. Rosales. 1992. Control of *Rhizoctonia* sheath blight and other diseases of rice by seed bacterization. In : *Biological control of plant diseases*. E.J. Tjamos et al. (eds.). Plenum Press, pp. 113-123.

Mew, T.W., A.K.W. Shahajahan and V. Mariappan. 1986. Diseases and disease management of rainfed low land rice. In : *Progress in rainfed lowland rice*. *Int. Rice Res. Inst.* P.O. Box 933. Manila, Philippines. pp. 339-348.

Mgonja, A.P. 1983. Sclerotia density and distribution of *Rhizoctonia solani* and their relationship to sheath blight of rice in Arkansas. Ph.D. Thesis, Univ. of Arkansas, p. 123.

Misawa, T. 1965. Nutritional factors for phytopathogenic fungi on culture media. *Ann. Phytopathol. Soc. Japan*, **31** : 27-34.

Miyake, I. 1910. Studien uber die Pilze der Reispflanze in Japan. *J. Coll. Agric. Imp. Univ. Tokyo*, **2** : 237-276.

Mizuta, H. 1956. On the relation between yield and inoculation times of sheath-blight, *Corticium sasakii* in the earlier planted paddy rice. *Asso. Pl. Prot. Kyushu*, **2** : 100-102.

Mochizuki, H., F. Araki and K. Yabutani. 1986. Rice sheath blight control with flutolanil. *J. Pesticide Sci.*, **12** : 29-33.

Mordue, J.E.M. 1974. *Thanatephorus cucumeris*. CMI Description of pathogenic fungi and bacteria No. **406**. CAB Intl., Lond., p. 2.

Moromizato, Z., N. Matsuyama and S. Wakimoto. 1980a,b. Effect of amino acids on sclerotium formation of *Rhizoctonia solani* Kühn (AG-1). I. Inhibition of sclerotial formation by various amino acids. II. Development process of sclerotium and its inhibition with several amino acids. *Ann. Phytopathol. Soc. Japan*, **46** : 15-20; 21-25.

Mrinalini, C. and D. Lalithakumari. 1995: Protoplast fusion : A biotechnological tool for strain improvement of *Trichoderma spp*. In : *Recent developments in biocontrol of plant pathogens* (K. Manibhushanrao and A. Mahadevan eds.), Today and Tomorrow's Printers and Publishers, New Delhi (In press).

Munnera, U.K. 1973. Studies on the control of sheath blight of rice. M.Sc. (Ag.) Thesis, Kerala Agric. Univ. Trivandrum, India, p. 85.

Murata, A. and J. Nagano. 1985. Rice sheath blight injury related with seedling density in a direct sowing system. *Proc. Kanto tosan Pl. Prot. Soc.*, **32** : 24-25.

Naiki, S. and T.T. Konoh. 1978. Indica rice varietal improvement for major diseases and insect resistance. *Pl. Prot. Bull., Taiwan*, **20** : 215-219.

Naicki, T. and T. Ui. 1975. Ultrastructure of sclerotia of *Rhizoctonia solani* Kühn invaded and decayed by soil micro-organisms. *Soil Biol. and Biochem.*, **7** : 301-304.

Nakanishi, T., Y. Takashi and K. Tamita, 1975. Mode of action of hemexazol. I. Effect of hemexazol on cell permeability of *Pellicularia sasakii*. *Ann. Phytopathol. Soc. Japan*, **41** : 321-328.

Namai, T., K. Hatakada and T. Asano. 1985. Identification of a bacterium which produces substances having antifungal activity against many important phytopathogenic fungi. *J. Agric. Res. Tohoku.*, **36** : 1-7.

Nandi, S. and N.K. Chakrabarthi. 1984. Differential virulence of isolates of *Corticium sasakii* the sheath blight pathogen to rice cultivars and the type of resistance offered by host plants. *Z. Pflkrankh. Pfschutz.*, **91** : 182-185.

Narayanaswamy, T. and A.V. Rao. 1984. *Waitea circinata* - a new fungus causing sheath blight of rice. *Curr. Sci.*, **53** : 874.

Nayak, P., K.V.S.R.K. Row and R. Sridhar. 1979. Host range of *Rhizoctonia solani*, the causal organism of sheath blight disease of rice. *Indian Phytopath.*, **32** : 604.

Newton, W. and N. Mayers. 1935. The physiology of *Rhizoctonia solani* Kühn III. The effect of toxic substances produced by *Rhizoctonia solani* Kühn when grown in liquid culture on the growth of wheat, carrots and turnips. *Sci. Agric.*, **15** : 393-401.

Nishikado, Y. and K. Hirata. 1937. Studies on the longevity of sclerotia of certain fungi under controlled environmental factors. *Ber. Ohara Inst.*, **7** : 535-547.

Nishikado, Y., K. Hirata and T. Higuti. 1938. Studies on the temperature relations to the longevity of pure culture of various fungi pathogenic to plants. *Ber. Ohara Inst.*, **8** : 107-124.

Nishimura, S. and M. Sasaki. 1963. Isolation of phytotoxic metabolites of *Pellicularia filamentosa*. *Ann. Phytopathol. Soc. Japan*, **28** : 228-238.

Nonaka, F. 1977. Causal fungus for lesion formation of various types on sheath of rice plant. *Proc. Assoc. Pl. Kyushu*, **23**: 15-17.

Nonaka, F., S. Ide and K. Tanaka. 1979. Cultural comparison of isolates of sheath blight fungus collected from different areas in Japan. *Proc. Assoc. Pl. Prot. Kyushu*, **25** : 1-4.

Nonaka, P. 1984. Sheath blight fungus *Rhizoctonia solani* Kühn (Sasakii type) : Its relation to sheath blight like diseases of rice plant. *Noyaku Kenkyu*, **31** : 14-20.

Nour El Dein, M.S. and M.S. Sharkas. 1965. Isolation of toxic substance from the culture filtrate of *R. solani*. *Phytopath. Z.*, **52** : 55-58.

Ogoshi, A. 1972. Grouping of *Rhizoctonia solani* Kühn with hyphal anastomosis. *Ann. Phytopathol. Soc. Japan.*, **38** : 117-122.

Ogoshi, A. 1975. Grouping of *Rhizoctonia solani* Kühn and their perfect stages. *Rev. Pl. Prot. Res.*, **8** : 93-103.

Ogoshi, A. 1976. Studies on the grouping of *Rhizoctonia solani* Kühn with hyphal anastomosis and on the perfect stages of groups. *Bull. Natn. Inst. Agric. Sci.*, **30** : 1-63.

Ogoshi, A., M. Ohiki, R. Sakai and T. Ui. 1979. Anastomosis grouping among isolates of binucleate *Rhizoctonia* *Trans. Mycol. Soc. Japan*, **20**: 33-39.

Ogura, H. and S. Akai. 1965. Studies on *Rhizoctonia solani* Kühn (*Pellicularia filamentosa* (Pat.) (Rogers). IV. The activity of antagonists to *Rhizoctonia solani* Kühn. *Ann. Phytopathol. Soc. Japan,* **30** : 219-224.

Ohata, K., K. Goto and T. Kozaka. 1966. Effect of low air temperature on the susceptibility of rice plants of blast disease, with special reference to some chemical components in the plants. *Bull. Natn. Inst. Agric. Sci. Japan,* **20** : 1-66.

Ohata, K. and K. Shimada. 1990. Forecasting of rice sheath blight disease by computer simulation model III. Effect of the incidence time on yield losses on rice. *Ann. Rep. Soc. Pl. Prot. North Japan,* **41** : 31-34.

Ohata, K. and K. Shimada. 1992. A computer simulation model for rice sheath blight disease. V. A forecasting for yield losses by disease. *Ann. Rep. Soc. Pl. Pro. North Japan,* **43** : 30-33.

Ohsumi, T., K. Tsushima, K. Maeda, S. Inoue and N. Itaya. 1989. Activity of bicyclic benzanilides against rice sheath blight. *J. Pestic. Sci.,* **14** : 229-233.

Okada, M. 1984. Control of moncelen (Pencycuron) for sheath blight disease in Miyazaki Prefecture. *Noyaku Kenkyu,* **31** : 29-32.

Okon, Y., I. Chat and Y. Hennis. 1973. Effect of lactose, ethanol and cycloheximide on the translocation pattern of radioactive compounds and sclerotium formation in *Sclerotium rolfsii. J. Gen. Microbiol.,* **74** : 251-258.

Ono, K. 1953. Damage of sheath blight to rice. *J. Pl. Prot.,* **7** : 99-103.

Ono, K. and K. Iwata. 1961. Control of sheath blight disease of rice plant by PCP. *Nogyo Oyobi Engei.,* **36** : 71-74.

Ou, S.H. 1976. Genetic defence of rice against disease. Symp. Genetic Basis of Epidemics in Agric., New York, April : 5-8.

Ou, S.H. 1985. *Rice diseases,* CAB International, Central Sales, Farnham Royal, Slough, U.K., p. 380.

Ou, S.H. and J.M. Bandong. 1976. Yield losses due to sheath blight of rice. *Int. Rice Res. Newsl.,* **1** : 14.

Ou, S.H., J.M. Bandong and F.L. Nuque. 1973. Some studies on sheath blight of rice at IRRI. *Int. Rice Res. Conf.,* Los Baños, Philippines, April : 23-27.

Ozawa, A. and T. Makino. 1990. Application of computerized forecasting system for rice sheath blight disease in Shizuoka Prefecture. *Proc. Kanto-Tosan Pl. Prot. Soc.,* **37** : 15-17.

Padmakumari, G. and S. Balakrishnan. 1987. Effect of organic amendments on the survival of *Rhizoctonia solani,* the sheath blight pathogen. In : *Proc. of workshop on biological control of plant diseases,* Tamil Nadu Agric. Univ., Coimbatore, March : 10-12.

Palo, M.A. 1926. *Rhizoctonia* disease of rice. I. A study of the disease and of the influence of certain conditions upon the viability of the sclerotial bodies of the causal fungus. *Philippine Agric.,* **15** : 361-375.

Papavizas, G.C. and C.B. Davey. 1960. *Rhizoctonia* disease of bean as affected by decomposing green plant materials and associated microflora. *Phytopathology,* **50** : 516-522.

Paracer, C.S. and D.S. Chahal. 1963. Sheath blight of rice caused by *Rhizoctonia solani* Kühn - A new record in India. *Curr. Sci.,* **32** : 328-329.

Park, M. and L.S. Bertus. 1932. Sclerotial disease of rice in Ceylon. I. *Rhizoctonia solani* Kühn. *Ceylon J. Sci.,* **11**: 319-331.

Parmeter, J.R.Jr. (ed.). 1970. *Rhizoctonia solani, Biology* and *Pathology,* Univ. of Calif. Press, Berkeley, Los Angeles and London, p. 225.

Parmeter, J.R.Jr. and H.S. Whitney. 1970. Taxonomy and nomenclature of the imperfect state. In : *Rhizoctonia solani Biology and Pathology.* Parmeter, J.R. Jr. (ed.) Univ. of Calif. Press, Berkeley, Los Angeles and London. pp. 7-19.

Parmeter, J.R.Jr., R.T. Sherwood and W.D. Platt. 1969. Anastomosis grouping among isolates of *Thanatephorus cucumeris. Phytopathology,* **59** : 1270-1278.

Peng, S.C., C.J. Tseng, C.L. Liu and T.P. Wang. 1965. The mycelial activities of rice sheath blight fungus in relation to the disease development. *Acta Phytophylac. Sin.,* **4** : 347-354.

Philip, R. and B.S. Joshi. 1980. *In vitro* toxicity of boron to *Corticium sasakii,* the incitant of sheath blight of rice. *Oryza,* **17** : 242-243.

Pillai, P.K. 1990. Laboratory screening of rice cultivars for sheath blight resistance. *Proc. Intl. Sym. Rice Res. : New Frontiers,* DRR. Hyderabad, pp. 237-239.

Pillai, P.K. 1993. Managing blast and sheath blight of rice in Haryana and Punjab. *Indian Farming,* 42 : 6-7.

Potieter, H.J. and M. Alexander. 1966. Susceptibility and resistance of several fungi to microbial lysis. *J. Bacteriol.*, **91** : 1526-1532.

Prabhath, C.A.M., M.R. Menon, L.R. Devi and C.K. Ramakrishnan. 1973. Varietal susceptibility of rice to infection by *Corticium sasakii* and its host range. *Agric. Res. J. Kerala*, **11** : 172-173.

Prabhath, C.A.M., M.R. Menon, L.R. Devi and C.K. Ramakrishnan. 1974. Studies on the viability of sclerotia of *Corticium sasakii* (Shirai) Matsumoto. *Agric. Res. J. Kerala*, **12** : 96-98.

Radha, A. 1991. Effect of mimosine on phytopathogenic fungi. M. Phil. Thesis, Univ. of Madras, Madras, p.85.

Rajan, C.P.D. 1987. Estimation of yield loss due to sheath blight of rice. *Indian Phytopath.*, **40** : 174-177.

Rajan, K.M. 1980. Soil amendments in plant disease control. *Int. Rice Res. Newsl.*, **5** : 515.

Rajan, K.M. 1981. Effect of major plant nutrients on the intensity of sheath blight and yield of grain and straw of rice. *Indian Potash J.*, **6** : 11-13.

Rajan, K.M. and S. Alexander. 1988. Mangement of sheath blight disease of rice with *Trichoderma viride* and some amendments in relation to population of pathogen in soil. *J. Biol. Control*, **2** : 36-41.

Rajan, K.M. and R.M. Menon. 1975. Effect of organic soil amendments on plant growth and intensity of sheath blight of rice. *Agric. Res. J. Kerala*, **13** : 179-181.

Ramalingam, P. 1983a. Studies on the phytotoxic metabolites of *Rhizoctonia solani* Kühn the causal organism of sheath blight disease of rice. Ph.D. Thesis, Univ. of Madras, Madras. p. 108.

Ramalingam, P. 1983b. Polygalacturonase production by *Rhizoctonia solani* Kühn, the causal organism of sheath blight disease of rice. *Curr. Sci.*, **52** : 371-372.

Ramalingam, P. 1984a. Effect of phytotoxic metabolites of *Rhizoctonia solani* Kühn on isoenzyme patterns of peroxidase. *Phytopathol. Mediter.*, **23** : 55-56.

Ramalingam, P. 1984b. Production of cellulolytic enzymes by *Rhizoctonia solani* Kühn. *Curr. Sci.*, **53**: 327-328.

Reddy, M.N. and A.S. Rao. 1975. Amino acids in mycelium and culture filtrates of *Rhizoctonia solani*. *Trans. Br. Mycol. Soc.*, **64** : 527-528.

Reddy, M.N. and A.S. Rao. 1978. Physiology of host-parasite relations in damping-off of groundnut caused by *Rhizoctonia solani*. *Phytopath. Z.*, **93** : 193-207.

Reinking, O.A. 1918. Philippine economic plant diseases. *Philippine J. Sci.*, **13** : 165-274.

Rogers, D.P. 1943. The genus *Pellicularia* (Thelephoraceae). *Farlowia.*, **1** : 95-118.

Rosales, A.M. and T.W. Mew. 1982. Antagonistic effects of soil microorganisms on rice sheath blight pathogen. *Intl. Rice Res. Newsl.* **7** : 12-13.

Rouse, D.I. and R. Baker,. 1978. Modeling and quantitative analyses of biological control mechanisms. *Phytopathology*, **68** : 1297-1302.

Roy, A.K. 1973. Natural occurrence of *Corticium sasakii* on some weeds. *Curr. Sci.*, **42** : 842-843.

Roy, A.K. 1976. Studies on the survival of sclerotia of *Corticium sasakii*. *Phytopath. Z.*, **86** : 270-275.

Roy, A.K. 1977a. Parasitic activity of *Trichoderma viride* on the sheath blight fungus (*Corticium sasakii*) of rice. *Z. Pflkrankh. Pflschutz.*, **84** : 675-683.

Roy, A.K. 1977b. Electrolytic changes in rice leaves infected with *Rhizoctonia solani*. *Indian Phytopath.*, **30** : 550-551.

Roy, A.K. 1978. Horizontal spread of sheath blight of rice plants in relation to spacing and nitrogen application. *Curr. Sci.*, **47** : 307-308.

Roy, A.K. 1979. Identification of strains of *Corticium sasakii* in India. *Int. Rice Res. Newsl.*, **4** : 17.

Roy, A.K. 1982. Susceptibility of rice plants to sheath blight at different stages. *Kavaka*, **7** : 25-26.

Roy, A.K. 1985a. Growth of sclerotia of *Rhizoctonia solani*. f. sp. *sasakii* in presence of soil inhabiting microorganisms. *Indian Phytopath.*, **38** : 45-49.

Roy, A.K. 1985b. Antagonistic effect of dhaincha on survival of *Rhizoctonia solani* f. sp. *sasakii*. *Int. Rice Res. Newsl.*, **10** : 9-10.

Roy, A.K. 1989. Source of seed-borne infection of sheath blight in rice. *Oryza*, **26** : 111-112.

Roy, A.K. 1991. Inhibitory effect of *Aspergillus terreus* on *Rhizoctonia solani* causing sheath blight of rice. *J. Indian. Bot. Soc.*, **70** : 95-98.

Roy, A.K. and U.N. Saikia. 1976. Chemical control of sheath blight of rice. *Indian Phytopath.*, **29** : 354-356.

Roy, A.K. and K.M. Sayre. 1984. Electron microscopical studies of *Trichoderma harzianum* and *T. viride* and mycoparasitic activity of the former on *Rhizoctonia solani* f. sp. *sasakii*. *Indian Phytopath.*, **37** : 710-712.

Rush, M.C. 1971. Factors contributing to yield losses in rice affected by the bordered leaf and sheath spot disease. In : 63rd *A. Prog. Rep. Rice Exp. Stn. Growley.*, I.A. pp. 221-225.

Rutger, R.J. 1990. Sheath blight resists skills of rice breeders. *Rice J.*, **93** : 23-25.

Ryker, T.C. 1939. The *Rhizoctonia* disease of Bermuda grass, sugarcane, rice and other grasses in Louisina. *Proc. Sixth Cong. Int. Soc. Sugarcane Tech. Baton Rouge*, 1938, pp. 198-201.

Ryker, T.C. and F.S. Gooch. 1938. *Rhizoctonia* sheath spot of rice. *Phytopathology,* **28** : 233-246.

Saikia, U.K. and A.K. Roy. 1975a. Pathogenicity of *Corticium sasakii* on some plants. *Indian Phytopath.*, **28** : 279-280.

Saikia, U.K. and A.K. Roy. 1975b. Leaf blight of turmeric caused by *Corticium sasakii*. *Indian Phytopath.*, **28** : 519-520.

Saksena, H.K. 1973. Banded blight disease of paddy in North India. In : *Intl. Rice Res. Conf.*, IRRI, Box 933, Manila, Philippines.

Saksena, H.K. and R.D. Chaubey. 1972. Banded blight disease of paddy. In : *All India Coordinated Rice Improvement Project* (AICRIP) *Workshop*, (presently D.R.R.), Hyderabad, India (Mimeographed).

Sanni, S.O. 1976. VA-mycorrhiza in some Nigerian soils. The effect of *Gigaspora gigantea* on the growth of rice. *New Phytol.*, **77** : 673-674.

Santos, L.G. 1970. Studies on the morphology, physiology and pathogenicity of *Corticium sasakii* (Shirai) Matsuomoto with special reference on the effects of nitrogen and carbon sources. M.S. Thesis, Coll. of Agric., Univ. of the Philippines, Philippines, Los Baños, Laguma, Philippines.

Sarkar, M.L. and A.K. Sinha. 1989. Use of phytoalexin inducing chemicals to control rice sheath blight (ShB). *Int. Rice Res. Newsl.*, **14** : 23.

Sasaki, S., N. Ota, J. Eguchi, Y. Furukawa, T. Akashiba, T. Tsuchiyama, S. Suzuki. 1969. Studies on polytoxins, antifungal antibiotics VIII. Mechanism of action on sheath blight of rice plant. *Ann. Phytopathol. Soc. Japan,* **34** : 272-279.

Sato, K. and T.S. Shoji. 1957. Pathogenicity of *Rhizoctonia solani* Kühn isolated from graminaceous weeds in forestry nurseries. *Bull. Fores. Exp. Stn., Meguro,* **96** : 89-104.

Sawada, K. 1912. On the 'shirakinibuyo' of camphor. *Spl. Bull. Formosan Agric. Exp. Stn.*, No. 4, p. 805.

Sha, X.Y. and L.H. Zhu. 1989. Resistance to sheath blight in China *Int. Rice Res. Newsl.*, **14** : 14-15.

Shanghai Institute of Agricultural Pesticides. 1975. Classification and identification of Jinggangmycin producing strains. *Acta Microbiol. Sin.*, **15** : 110-113.

Shanghai Institute of Plant Physiology. 1975. *Native method of production* and *utilization of 'Ching-feg-mei-su'*., p. 47.

Sharma, B.D. and S.K. Mukherji. 1978. Natural occurrence of *Corticium sasakii* on four hosts. *Sci. and Cul.*, **44** : 43-44.

Sharma, N.R. and P.S. Teng. 1990a,b. I. Effect of rice growth stage on sheath blight (ShB) development and yield loss. II. Comparison of rice sheath blight (ShB) assessment methods. *Int. Rice Res. Newsl.*, **15** : 19-20; 20-21.

Sherwood, R.T. 1969. Morphology and physiology in four anastomosis groups of *Thanatephorus cucumeris.* *Phytopathology,* **59** : 1924-1929.

Sherwood, R.T. 1970. Physiology of *Rhizoctonia*. In : *Rhizoctonia solani Biology* and *Pathology,* (ed.) J.R. Parmeter, Jr., Univ. of Calif. Press, Berkeley, Los Angeles and London, pp. 69-92.

Shigemoto, R., T. Okuno and K. Matsuura. 1989. Effect of validamycin A on the activity of trehalase of *Rhizoctonia solani* and several sclerotial fungi. *Ann. Phytopathol. Soc. Japan,* **55** : 238-241.

Shirai, M. 1906. On *Hypochnus sasakii* n. sp. *Bot. Magazine, Tokyo,* **20** : 319-323.

Shukla, R.P., R.K. Singh and R.S. Dwivedi. 1990. Toxicity of essential oils against *Rhizoctonia solani* Kühn fungus causing sheath blight (ShB) in rice. *Int. Rice Res. Newsl.*, **15** : 27.

Singh, P.A. and M.S. Pavgi. 1969. Oriental sheath and leaf spot of rice. *Pl. Dis. Reptr.*, **53** : 444-445.

Singh, U.D. and N. Sethunathan. 1987. Individual and combined effects of certain pesticides on *Rhizoctonia solani*, sheath blight pathogen of rice. *J. Phytopath.*, **119** : 240-247.

Sreenivasaprasad, S. 1988. Studies on the hyperparasitic and biocontrol potential of *Gliocladium virens* and *Trichoderma longibrachiatum*. Ph.D. Thesis, Uni. of Madras, Madras, p.122.

Sreenivasaprasad, S. and K. Manibhushanrao. 1990. Antagonistic potential of *Gliocladium virens* and *Trichoderma longibrachiatum* to phytopathogenic fungi. *Mycopathologia*, **109** : 19-26.

Sridhar, R., K. Manibhushanrao and A.K. Sinha. 1989. Innovative approaches to management of some major diseases of rice. *J. Sci. and Ind. Res.*, **48** : 181-185.

Sumina, I., T. Kitamura, T. Baba and Y. Usui. 1966. Controlling effect of halogenophenyl pyrozolone and halogenophenyl hydrazine derivatives on some plant diseases. *Ann. Phytopathol. Soc. Japan*, **32** : 175-180.

Suryadi, Y. and T.S. Kadir. 1989. Field evaluation of fungicides to control rice sheath blight (ShB). *Int. Rice Newsl.*, **14** : 35.

Takahi, Y., T. Nakanishi and K. Nakagani. 1974. Studies on hymexazol (3-hydroxy-5-methylisoxazole, Tachigaran) on the effectiveness and characteristics in rice sheath blight control. *Proc. Kansai Pl. Prot. Soc.*, **16** : 52-58.

Takahi, Y. and seven others. 1988. Development of Diclomezine, a new fungicide for control of rice sclerotial diseases caused by *Rhizoctonia* and *Sclerotium*. In : Symposium IV, Development of new fungicides active against basidiomycetes. 5th *Int. Congress Pl. Path., Kyoto, Japan*, p. 23 (Abstr.).

Takatsu, S. and J. Nishimura. 1962. Studies on the control of sheath blight of rice plant. 4. Effects of sodium salt of pentachlorophenol. *Bull. Hyogo Pref. Agric. Exp. Stn.*, No. **10** : 27-30.

Takita, K., T. Tanaka, K. Nose and K. Fukunaga. 1965. Organo-arsenic fungicide : Relationship between antimicrobial activity and chemical constitution of organo-arsenic compounds. *Bull. Natl. Inst. Agric. Sci. Tokyo, Ser. C.*, No. **19** : 1-46.

Talbot, P.H.B. 1965. Studies of *Pellicularia* and associated genera of Hymenomycetes. *Persoonia.*, **3** : 371-406.

Talbot, P.H.B. 1970. Taxonomy and nomenclature of the perfect state. In : *Rhizoctonia solani, Biology* and *Pathology*, ed. J.R. Parmeter Jr., Univ. of Calif. Press, Berkeley and London, pp. 20-31.

Tamura, H. 1965. Effect of organotin fungicides applied for the control of rice blast, and rice sheath blight. *Bull. Natl. Inst. Agric. Sci.*, No. **19** : 47-79.

Tazaki, H., N. Matsuyama and S. Wakimoto. 1982. Effect of sulfur-containing amino acids on sclerotial formation of *Sclerotium rolfsii*. *Proc. Assoc. Pl. Prot. Kyushu*, **28** : 69-70.

Templetion, G.E. and T.H. Johnson. 1969. Brown-bordered leaf and sheath spots on rice. *Ark. Farm. Res.*, **18** : 5.

Tervet, I.W. 1937. An experimental study of some fungi injurious to seedling flax. *Phytopathology*, **27** : 531-546.

Tewari, S.N. and A.P. Dath. 1984. Effect of leaf extract media of some plants on the growth of three fungal pathogens of rice. *Indian Phytopath.*, **37** : 458-461.

Tewari, S.N., and N. Mandakini. 1991. Activity of four plant leaf extracts against the three fungal pathogens of rice. *Trop. Agric.*, **68** : 373-375.

Thakur, R.S., S.K. Sugha and B.M. Singh. 1992. Anastomosis grouping of isolates of *Rhizoctonia solani*. *Indian Phytopath.*, **45** : 109-110.

Thara, K.V., and S.S. Gnanamanickam, 1994. Biological control of rice sheath blight in India: lack of correlation between chitinase and production by bacterial antagonists and sheath blight suppression. *Plant Soil*, **160**: 227-280.

Thirumalachar, M.J., M.S. Pavgi and R.A. Singh. 1969. *In vitro* activity of three antifungal antibiotics against *Protomyces macrosporus*, *Taphrina maculans* and *Corticium sasakii*. *Hindustan Antibiotics Bull.*, **11** : 189-190.

Tsai, W.H. 1970. Studies on the relation between rice blast diseases. 1. Observations on the host range of rice sheath blight fungus, *Pellicularia sasakii* on weeds. *J. Taiwan Agric. Res.*, **19** : 48-50.

Tsai, W.H. 1973. Variation in cultural characteristics and pathogenicity of single basidiospores of *Thanatephorus cucumeris* (Frank) Donk, causing sheath blight of rice. M.S. Thesis, Univ. of the Philippines, Los Baños, Laguna, Philippines, p. 85.

Tsai. W.H. 1976. The influence of rice sheath blight on yield at different inoculation stages. *Ann. Rep. Taiwan Agric. Res. Inst.*, 1975, pp. 115-116.

Tu, C.C. and J.W. Kimbrouogh. 1978. Systematics and phylogeny of fungi in the *Rhizoctonia* complex. *Bot. Gazette.*, **139** : 454-466.

Tu, C.C., Y.C., Chang and W.C. Wang. 1979. Studies on ecology of *Rhizoctonia solani*, the causal organism of sheath blight of rice. *Natl. Sci. Coun.*, **7** : 1208-1219.

Tu, J.C. 1967. Strain of *Pellicularia sasakii* isolated from rice in Taiwan. *Pl. Dis. Reptr.*, **51** : 682-684.

Tu, J.C. 1968. Physiological specialization of strains of *Pellicularia sasakii* isolated from rice plants. *Pl. Dis. Reptr.*, **52** : 323-326.

Ui, T., T. Naiki and M. Akimoto. 1976. A seiving-floatation technique using hydrogen peroxide solution for determination of sclerotial populaiton of *Rhizoctonia solani* Kühn in soil. *Ann. Phytopathol. Soc. Japan*, **42** : 46-48.

Van der Plank, J.E. 1968. *Disease resistance in plants*. Academic Press, New York, p. 206.

Valdez, R.B. 1955. Sheath spot of rice. *Philipp. Agric.*, **33** : 317-337.

Vanetten, H. and D.F. Bateman. 1965. Proteolytic activity in extracts of *Rhizoctonia* infected hypocotyls of bean. *Phytopathology*, **55** : 1285.

Varma, A.S. and M.R. Menon. 1977. Fungicidal trial on the control of sheath blight of rice. *Madras Agric. J.*, **64** : 416-417.

Varma, A.S., C.K. Peethambaram, S. Balakrishnan and M.R. Menon. 1980. *In vitro* effects of certain herbicide formulations on *Corticium sasakii*. *Agric. Res. J. Kerala*, **16** : 114-116.

Vasanthadevi, T., R. Malarvizhi, N. Sakthivel and S.S. Gnanamanickam. 1989. Biological control of sheath blight of rice in India with antogonistic bacteria. *Pl.* and *Soil*, **119** : 325-330.

Vasavan, M.C., K.M. Rajan and T. Thomas. 1980. Herbicides in plant disease control. *Intl. Rice Res. Newsl.*, 5 : 4.

Venkataraman, S., C. Challam and A.P.K. Reddy. 1995. Saprophytic phase of *Rhizoctonia solani* and its biocontrol. In : *Recent developments in biocontrol of plant pathogens* (K. Manibhushanrao and A. Mahadevan eds.), Today and Tomorrow's Printers and Publishers, New Delhi (In press).

Venkatnarayan, S.V. 1949. The validity of the name *Pellicularia koleroga* Cooke. *Indian Phytopath.*, 2 : 186-189.

Vijayan, M. and M.C. Nair. 1981. Anastomosis grouping of isolates of *Rhizoctonia solani* Kühn (*Thanatephorus cucumeris* Frank Donk), causing sheath blight of rice. *Curr. Sci.*, **54** : 289-291.

Vidyasekaran, P. and M. Muthamilan. 1995. Development of formulations of *Pseudomonas fluoresence* for control of check-pea wilt. *Plant Disease* (In press).

Vidyasekaran, P., M. Muthamilan, R. Rabindran, K. Sethuraman and C.N. Ananthakumar. 1995. Development of a powdery formulations to control root and foliar pathogens. In : *Recent developments in biocontrol of plant pathogens* (K. Manibhushanrao and A. Mahadevan eds.), Today and Tomorrow's Printers and Publishers, New Delhi (In press).

Viji, G. 1991. Studies on the biocontrol agents *Gliocladium* and *Trichoderma* spp. M.Phil. Thesis, Univ. of Madras, Madras. p. 40.

Viji, G., U.I. Baby and K. Manibhushanrao. 1993. Induction of fungicidal resistance in *Trichoderma* spp. through UV irradiation. *Indian J. Microbiology*, **33** : 125-129.

Vilgalys, R. 1988. Genetic relatedness among anastomosis groups in *Rhizoctonia solani*, as measured by DNA/DNA hybridization. *Phytopathology*, **78** : 698-702.

Vilgalys, R. and D. Gonzalez. 1990. Ribosomal DNA restriction fragment length polymorphism in *Rhizoctonia solani*. *Phytopathology*, **80** : 151-158.

Waheeta, A., S. Sreenivasaprasad and K. Manibhushanrao. 1987. Induced resistance in rice to sheath blight disease. *Curr. Sci.*, **56** : 486-489.

Wakae, O. and K. Matsuura. 1975. Characteristics of validamycin as a fungicide for *Rhizoctonia* disease control. *Rev. Pl. Prot. Res.*, **8** : 81-92.

Watanabe, B. and A. Matsuda. 1966. Appointed experiment, plant disease and insect pests. *Bull. No. 7 Agric. For Fish. Res. Coun. Japan*, Tokyo.

Wei, C.T. 1934. *Rhizoctonia* sheath blight of rice. *Bull. Coll. Agric. Fores. Univ. Nanking.*, No. **15**, p. 21.

Wickemasingha, D.L. and Y.J.P.K. Mitrasena. 1988. Sheath blight disease of rice in the low country wet zone of Sri Lanka : Effect on yield and factors influencing spread and distribution (Abstr.) In : Int. symp. on crop losses due to disease outbreaks in the tropics and counter measures. *Min. Agric., For. Fish., Japan*, pp. 11.

Wu, Y.L. 1971. Varietal differences in sheath blight resistance of rice obtained in Southern Taiwan. *Sabro Newsl.*, **3** : 1-5.

Yahara, M., N. Matsuyama and S. Wakimoto. 1985. Inhibition of sclerctial formation of *Rhizoctonia solani* with sulfur containing substances II. *Proc. Assoc. Pl. Prot. Kyushu,* **31** : 226.

Yamada. 1986. Moncern (pencycuron), a new fungicide. *Agric. Chem. Inst. Nehon Tokushu, Nyoku Seizo K.K.* p 26.

Yamaguchi, T., K. Iwata and T. Kuramoto. 1971. Study on forecasting of the sheath blight of rice plant caused by *Pellicularia sasakii* (Shirai) S. Ito. I-II. *Bull. Hokuriku Natl. Agric. Exp. Stn.,* **13** : 15-34; 35-37.

Yamaguchi, T., K. Kardin, Wagiman and D.M. Tantera. 1982. Hyphal anastomosis groups of *Rhizoctonia solani* Kühn in Indonesia. *Res. Rep. Japan-Indonesia Agric. Res. proj.,* pp. 123-126.

Yamamoto, H. 1985. Development of validamycin, its controlling effect on rice sheath blight. *Japan Pestic. Inf.,* **47** : 17-22.

Yamamoto, H., T. Iwata, M. Shibata, K. Mizumo and A. Miyake. 1965. *Streptomyces multisporalis* sp. nov. and a new antibiotic neohumidin, which controls sheath blight, *Pellicularia filamentosa* f. *sasakii* on rice. *Agric. Biol. Chem.,* **29** : 360-368.

Yarwood, C.E. 1967. Response to parasites. *Ann. Rev. Pl. Physiol.,* **18** : 419-438.

Yokogi, K. 1927. On the *Hypochnus* disease of soyabeans and its comparison with that of rice plants. *J. Pl. Prot.,* **14** : 12.

Yokoyama, T. and K. Okuhara. 1990. Chemical control of rice sheath blight based on relation between percentage of the lesions and grain quality of rice. *Proc. Assoc. Pl. Prot. Kyushu,* **36** : 8-10.

Yokoyama, T. and K. Okuhara. 1992. Severity and yield loss by rice sheath blight disease on early maturing-variety of rice (Variety Natsuhikari). *Kyushu Agric. Res.,* **54** : 79.

Yoshimura, D. 1984. Chemical control of sheath blight. *Noyaku Kenkyu,* **31** : 23-28.

Yoshimura, S. 1954. On the scale for estimating degree of severity of sheath blight by *Hypochnus sasakii* Shirai in rice plant. *Ann. Phytopathol. Soc. Japan,* **19** : 58-60.

Yoshimura, S. 1955a. Nutritional and physiological studies on the rice sheath blight fungus (*Hypochnus sasakii* Shirai). I. On the cellulose decomposition by the rice sheath blight fungus. *Kyushu Agric. Res.,* **15** : 62-64.

Yoshimura, S. 1955b. On the effect of the shading upon the susceptibility of the rice plant to the sheath blight, *Hypochnus sasakii* Shirai. *Kyushu Agric. Res.,* **16** : 113.

Yoshimura, S. and T. Nishizawa. 1954. Studies on the method of testing varietal resistance of upland rice plants caused by sheath blight, *Hypochnus sasakii* Shirai. *Bull. Kyushu Agric. Exp. Stn.,* No. **2** : 361-376.

Yu, C.M. 1975. Effect of nutritional and microclimatic conditions on the development of sheath blight disease of rice. M.S. Thesis, Univ. of the Philippines, Los Baños, Laguna, Philippines, p. 48.

Yu, C.M., Y.T. Huang and H.J. Tsay. 1980. Disease development of rice sheath blight. *Pl. Prot. Bull. Taiwan,* **22** : 263-267.

Yu, C.M., K.C. Ling and S.H. Ou. 1976. Effect of nutritional and microclimate conditions on the development of sheath blight disease of rice. *Pl. Prot. Bull. Taiwan,* **18** : 261-267.

Yu, S.H. 1977. Ecological studies on the transition of sheath blight of rice in Korea. *Res. Rep. Inst. Agric. Sci. Tech. Coll. Agric. Chungnam Natl. Univ.,* **4** : 283-316.

Yu, S.H. 1981. Occurrence of rice sheath blight and its control in Korea. *Korean J. Pl. Prot.,* **20** : 59-66.

Zuber, M. and K. Manibhushanrao. 1979. Phytoalexin from the germinating seed of rice (*Oryza sativa* L.). *Curr. Sci.,* **49** : 497-498.

Zuber, M. and K. Manibhushanrao. 1982. Studies on comparative gel electrophoretic patterns of proteins and enzymes from isolates of *Rhizoctonia solani* causing sheath blight disease in rice. *Can. J. Microbiol.,* **28** : 762-771.

Zuber, M. and K. Manibhushanrao. 1983. Physiology of host-pathogen interaction in rice-sheath blight disease with reference to changes in carbohydrates and nitrogen contents. *Indian J. Bot.,* **6** : 117-124.

Zuber, M. and K. Manibhushanrao. 1984. Changes in phenolics and enzymes of phenol metabolism in sheath blight disease of rice caused by *Rhizoctonia solani*. *Acta Phytopath. Hung.,* **19** : 251-262.

Zuber, M., N. Matsuyama and S. Wakimoto. 1984. Inhibition of sclerotial formation of *Rhizoctonia solani* with sulfur-containing substances-I. *Proc. Assoc. Pl. Prot. Kyushu,* **30** 176.

Index

[Note : Rice cultivars and ShB pathogen *Rhizoctonia. solani* are not listed]